Student Book

Series Editors: *Trevor Johnson & Tony Clough*

Aim High 1
Aiming for Grade C
in Edexcel GCSE Mathematics

Linear and Modular

A PEARSON COMPANY

Contents

NUMBER AND ALGEBRA

Contents

HANDLING DATA

Introduction

Welcome to *Aim High* – the grade-boosting revision guide for exam success.

Edexcel has studied the performance in exams of hundreds of thousands of students to find out which topics students find most difficult and the common errors they make. Written by examiners and based on Edexcel's own results data, *Aim High* targets all the areas where students regularly lose most marks in the exams. It helps you prioritise the skills and techniques on which to focus your revision, how to avoid losing marks and how to gain marks through showing your workings.

Aim High 1 is designed to help you achieve a Grade C. It is full of advice, hints and tips so you gain as many marks as possible from your working and answers. By focusing your revision on these skills and techniques, you improve your chance of getting the grade you need.

After this introduction is some important general exam advice and exam hints. The key maths topics are covered on pages 10 to 116. All the topics in the book have been chosen because students regularly lose marks in these areas and your teacher will help you decide which skills you should focus on. You can use the colour-coding to help you plan your revision programme: full steam ahead now with the essential green topics; get ready too for the important amber topics; and, finally, make sure your red skills don't hold you back. At the end of the book you will find a list of key terms and exam vocabulary, and answers to the questions.

Features of an *Aim High* topic

You may not need to practise all the topics in this book. Your teacher will tell you which ones to focus on. Tackle the important green topics first. Then work through the amber topics. If you still have time, work on the red topics too.

These are the maths skills you need to practise in order to succeed at the topic.

Each unit starts with a reminder of what you need to know on this topic. Make sure you can remember these facts.

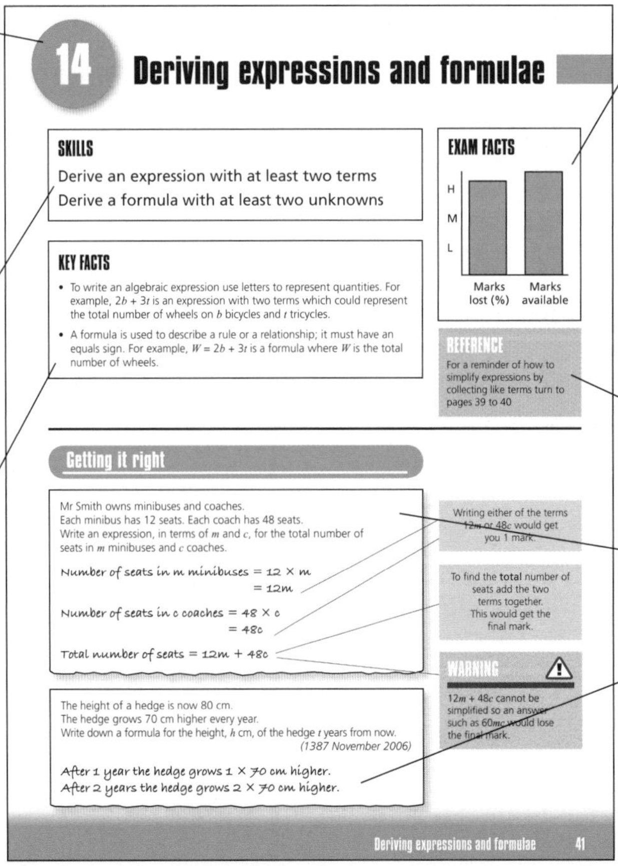

The first bar in the bar chart shows how students perform when tested: the higher the bar the worse students perform at these skills. The second bar shows how many marks are available in the exams from questions like these. In this example, a lot of marks are available *and* students lose a high percentage of those marks.

Other sections in the *Aim High 1* book on this topic which you may find helpful.

Questions similar to those you may come across in the exam.

You can gain marks from your workings even if your final answer is wrong. The worked examples have been written to show you how to gain maximum marks from your answers.

14 Deriving expressions and formulae

SKILLS

Derive an expression with at least two terms
Derive a formula with at least two unknowns

KEY FACTS

- To write an algebraic expression use letters to represent quantities. For example, $2b + 3t$ is an expression with two terms which could represent the total number of wheels on b bicycles and t tricycles.
- A formula is used to describe a rule or a relationship; it must have an equals sign. For example, $W = 2b + 3t$ is a formula where W is the total number of wheels.

EXAM FACTS

Marks lost (%)   Marks available

REFERENCE

For a reminder of how to simplify expressions by collecting like terms turn to pages 39 to 40

Getting it right

Mr Smith owns minibuses and coaches.
Each minibus has 12 seats. Each coach has 48 seats.
Write an expression, in terms of m and c, for the total number of seats in m minibuses and c coaches.

Number of seats in m minibuses $= 12 \times m$
$= 12m$

Number of seats in c coaches $= 48 \times c$
$= 48c$

Total number of seats $= 12m + 48c$

Writing either of the terms $12m$ or $48c$ would get you 1 mark.

To find the **total** number of seats add the two terms together. This would get the final mark.

WARNING ⚠

$12m + 48c$ cannot be simplified so an answer such as $60mc$ would lose the final mark.

The height of a hedge is now 80 cm.
The hedge grows 70 cm higher every year.
Write down a formula for the height, h cm, of the hedge t years from now.
(1387 November 2006)

After 1 year the hedge grows 1×70 cm higher.
After 2 years the hedge grows 2×70 cm higher.

Deriving expressions and formulae   41

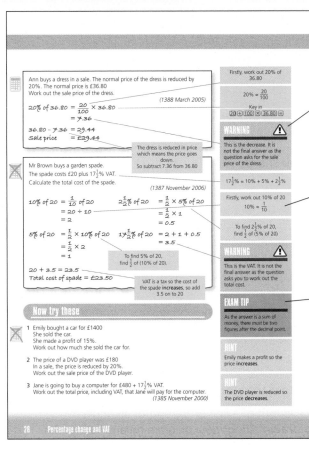

Ann buys a dress in a sale. The normal price of the dress is reduced by 20%. The normal price is £36.80
Work out the sale price of the dress.

(1388 March 2005)

$20\% \text{ of } 36.80 = \frac{20}{100} \times 36.80$

$= 7.36$

$36.80 - 7.36 = 29.44$

Sale price $= £29.44$

The dress is reduced in price which means the price goes down. So subtract 7.36 from 36.80

Firstly, work out 20% of 36.80

$20\% = \frac{20}{100}$

Key in
20 ÷ 100 × 36.80 =

WARNING ⚠

This is the decrease. It is not the final answer as the question asks for the sale price of the dress.

Mr Brown buys a garden spade.
The spade costs £20 plus $17\frac{1}{2}\%$ VAT.
Calculate the total cost of the spade.

(1387 November 2006)

$10\% \text{ of } 20 = \frac{1}{10} \text{ of } 20$
$= 20 \div 10$
$= 2$

$2\frac{1}{2}\% \text{ of } 20 = \frac{1}{2} \times 5\% \text{ of } 20$
$= \frac{1}{2} \times 1$
$= 0.5$

$5\% \text{ of } 20 = \frac{1}{2} \times 10\% \text{ of } 20$
$= \frac{1}{2} \times 2$
$= 1$

$17\frac{1}{2}\% \text{ of } 20 = 2 + 1 + 0.5$
$= 3.5$

To find 5% of 20, find $\frac{1}{2}$ of (10% of 20).

$20 + 3.5 = 23.5$
Total cost of spade $= £23.50$

VAT is a tax so the cost of the spade **increases**, so add 3.5 on to 20

$17\frac{1}{2}\% = 10\% + 5\% + 2\frac{1}{2}\%$

Firstly, work out 10% of 20
$10\% = \frac{1}{10}$

To find $2\frac{1}{2}\%$ of 20, find $\frac{1}{2}$ of (5% of 20)

WARNING ⚠

This is the VAT. It is not the final answer as the question asks you to work out the total cost.

EXAM TIP

As the answer is a sum of money, there must be two figures after the decimal point.

Now try these

1 Emily bought a car for £1400
She sold the car.
She made a profit of 15%.
Work out how much she sold the car for.

2 The price of a DVD player was £180
In a sale, the price is reduced by 20%.
Work out the sale price of the DVD player.

3 Jane is going to buy a computer for £480 + $17\frac{1}{2}\%$ VAT.
Work out the total price, including VAT, that Jane will pay for the computer.
(1385 November 2000)

HINT

Emily makes a profit so the price **increases**.

HINT

The DVD player is reduced so the price **decreases**.

Avoid common mistakes.

Advice on how to pick up marks and how to avoid losing marks.

General tips on good habits to help you stay on track.

Questions give you a chance to practise the skills. The questions in these exercises progress from easy to more difficult.

If you need help to get started, the hints tell you how to tackle the question.

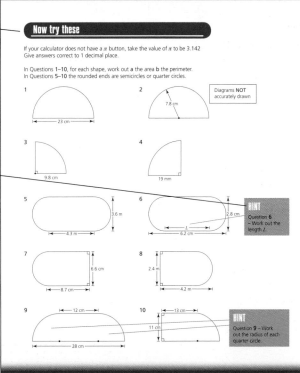

Now try these

If your calculator does not have a π button, take the value of π to be 3.142
Give answers correct to 1 decimal place.

In Questions 1–10, for each shape, work out **a** the area **b** the perimeter.
In Questions 5–10 the rounded ends are semicircles or quarter circles.

1

2

Diagrams **NOT** accurately drawn

7.8 cm

23 cm

3

4

9.8 cm

19 mm

5

3.6 m

4.3 m

6

2.8 cm

L

6.2 cm

HINT

Question **6** – Work out the length L.

7

6.6 cm

8.7 cm

8

2.4 m

4.2 m

9

12 cm

11 cm

28 cm

10

13 cm

HINT

Question **9** – Work out the radius of each quarter circle.

Good luck with your revision!

Exam advice

- **Make sure you have all the necessary equipment.**

  This includes a 30 cm ruler, a protractor, a pair of compasses and, when allowed, a calculator.

- **Write in black or blue ink.**

  Don't use pencil or fluorescent ink. Now that scripts are marked online, it is vital that answers can be read by a scanner.

- **For diagrams use an HB pencil, but it should not be too sharp.**

  For scanning purposes, a harder pencil or a very sharp one does not show up well on a grid.

- **Show working in the space provided for each question.**

  Don't go outside the working space allowed for each question and don't write on the formulae page or on blank pages. If necessary, ask for a supplementary answer sheet and use that.

- **Don't alter your working – cross it out and replace it.**

  If you realise you have made a mistake, cross out the error with a single line and replace it with the correct working.

- **Don't give the marker a choice of answers or methods.**

- **Don't take measurements from a diagram, if you are told that it is not accurately drawn.**

  Sometimes a question specifically instructs you to find the length of a line or the size of an angle by measuring but the statement *Diagram NOT accurately drawn* alongside a diagram means that taking measurements from the diagram will not give correct answers.

- **Tracing paper is useful for transformations.**

  In the examination, you can ask for tracing paper, which may help you answer questions on reflections and rotations.

Exam hints

- **Show all your working.**

Example	Notes

Q   Solve $5x - 7 = 2x - 5$
 	
 A $5x - 2x = 7 - 5$ ✓
        $3x = 2$ ✓
          $x = 1\frac{1}{2}$ ✗
	An incorrect answer with no working can score no marks. An incorrect answer with correct working will often receive the majority of the marks. The attempt in the example would score 2 marks out of 3 but, without the working, it would score no marks.

- Before rounding, show more figures than the question asks for.

Example	Notes
Q Find the circumference of a circle with a diameter of 9.5 cm. Give your answer correct to 1 decimal place. A $\pi \times 9.5 = 29.8451\ldots$ ✓ Circumference = 29.9 cm ✗	$29.8451\ldots$ shows that the correct calculation was keyed into the calculator and that the error occurs at the rounding stage.

- Make a rough estimate of calculations.

Example	Notes

Q   Find the area of a circle with a radius of 4.2 cm. Give your answer correct to 1 decimal place.
 	
 A $\pi \times 4.2^2$ ✓
       $= 26.38\ldots\ldots$ ✗
    Area $= 26.4$ cm^2 ✗
    Check: A rough estimate is 3×4^2
    $3 \times 4^2 = 3 \times 16 = 48$
	The usual method for finding rough estimates is to round each number to 1 significant figure.
	 A rough estimate does not tell you whether your answer is right but it does tell you whether it is reasonable; in this case, it is not.

- Whenever possible, ask yourself 'Is my answer sensible?'

Example	Notes

Q   Work out the value of x. Give your answer correct to 1 decimal place.
 	
    8 cm x cm   5 cm
 	
 A $x^2 = 8^2 + 5^2$ ✗
      $= 64 + 25 = 89$ ✗
    $x = \sqrt{89} = 9.4$ ✗
	This answer is **not** sensible as, in a right-angled triangle, the longest side is the hypotenuse, the one opposite the right angle.
	 (The first line of working should be $8^2 = x^2 + 5^2$)

- Whenever possible, check your answers.

Example	Notes

Q $v = u + at$
    Find a when $v = 4$, $u = 10$ and $t = 2$
 	
 A $4 = 10 + 2a$ ✓
    $2a = 4 - 10 = -6$ ✓
    $a = -3$ ✓
    Check by substituting the values of u, a, and t into $v = u + at$ to find the value of v.
    $10 - 3 \times 2 = 10 - 6 = 4$ ✓
	There are many ways of checking answers. When solving equations, for example, check that your solution fits the original equation. It is sometimes possible, as a check, to work a calculation backwards.

Writing a number as a product of its prime factors

SKILLS

Write a number as a product of its prime factors

Write a number as a product of powers of its prime factors

KEY FACTS

- All whole numbers, except 1, are either prime or can be written as a product of prime numbers.
- The first five prime numbers are 2, 3, 5, 7 and 11

EXAM FACTS

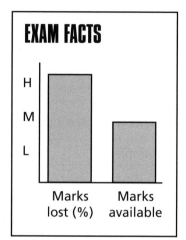

Getting it right

Write 120 as a product of its prime factors.

(1388 March 2003)

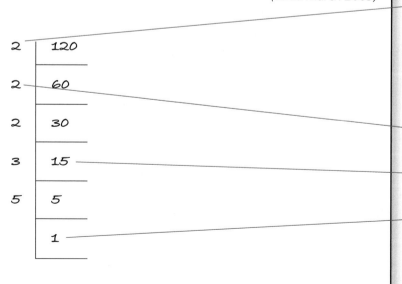

Look for a small prime number that is a factor of 120
120 = 2 × 60
Write the 2 next to the 120, and the 60 beneath.

Now look for a small prime number that is a factor of 60
60 = 2 × 30

15 is an odd number, so 2 is not a factor of 15

Stop at 1

WARNING

A common error is to forget to write down the answer.

Questions can also be answered by using a factor tree.

Write 495 as
i   a product of its prime factors,
ii a product of powers of its prime factors.

i

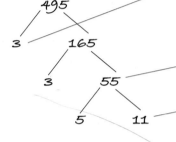

$$495 = 3 \times 3 \times 5 \times 11$$

ii $495 = 3^2 \times 5 \times 11$

The smallest prime factor of
495 is 3
$495 = 3 \times 165$

5 is also a factor of 495. You
could start this factor tree
with $495 = 5 \times 99$

$165 = 3 \times 55$

Both 5 and 11 are prime
numbers.

WARNING

Remember to write down
the product of the prime
numbers at the end of each
branch.

3×3, written as a power of
3, is $= 3^2$

Now try these

In Questions **1–8**, write each number as a product of its prime factors.

1 40   **2** 45   **3** 52   **4** 54   **5** 72

6 96   *(1387 June 2003)*

7 140   *(1388 March 2005)*

8 108   *(1388 June 2006)*

In Questions **9** and **10**, write each number as a product of powers of its prime factors.

9 200

10 1000

2 Highest common factor

SKILL

Find the highest common factor (HCF) of two or more whole numbers

EXAM FACTS

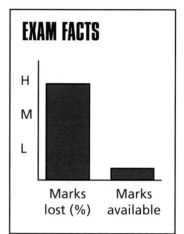

KEY FACTS

- A factor of a number **divides** exactly into that number. For example, 5 is a factor of 15

- If two numbers have the same factor, it is called a **common factor**. For example, 1, 2 and 4 are **common factors** of 20 and 24 as they are factors of both 20 and 24

- The **highest common factor** (HCF) of two or more numbers is the largest number which is a factor of those numbers. For example, the highest common factor (HCF) of 20 and 24 is 4

Getting it right

List all the factors of 24

1 × 24
2 × 12
3 × 8
4 × 6

6 × 4

The factors of 24 are
1, 2, 3, 4, 6, 8, 12, 24

WARNING

A common error is to omit 1 and the number itself from the list of factors.

List the factors in pairs to make sure that all are included. Start with 1, then try 2, 3 and so on.

Really the same as 4 × 6 So you know you can stop at 4 × 6

Find the highest common factor (HCF) of 64 and 88

The factors of 64 are
1, 2, 4, 8, 16, 32, 64

The factors of 88 are
1, 2, 4, 8, 11, 22, 44, 88

HCF = 8

List all the factors of 64

List all the factors of 88

Find the highest number which appears in both lists. This is the highest common factor.

Now try these

From the numbers in the ring,

1 write down a factor of 36,
2 write down a number which has 10 as a factor.

12
8   16
20   24

In Questions **3–7**, list all the factors of the numbers.

3 12   **4** 18   **5** 20   **6** 36   **7** 96

In Questions **8–12**, find the highest common factor (HCF) of the two numbers.

8 12, 18   **9** 24, 36   **10** 42, 56   **11** 72, 84   **12** 96, 120

In Questions **13–17**, find the highest common factor (HCF) of the three numbers.

13 8, 10, 12   **14** 8, 12, 16   **15** 9, 12, 18   **16** 12, 18, 30   **17** 30, 36, 45

18 Jonathon says '2 is always a common factor of any pair of even numbers.'
Is Jonathon correct?
You must give a reason.

19 Becky says '2 is always the highest common factor of any pair of even numbers.'
Is Becky correct?
You must give a reason.

20 a Find two **odd** numbers which have 15 as their highest common factor.
b Is it possible to find two **even** numbers which have 15 as their highest common factor? Explain your answer.

3 Lowest common multiple

SKILL

Find the lowest common multiple (LCM) of two or more whole numbers

EXAM FACTS

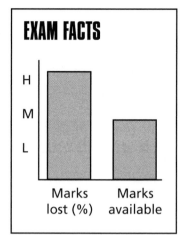

KEY FACTS

- A **multiple** of a number is found by multiplying it by any whole number. For example 3, 6 and 60 are multiples of 3

- If a number is a multiple of two or more given numbers, it is called a **common multiple**. For example 12 is a common multiple of 4 and 6

- The **lowest common multiple** (LCM) of two or more numbers is the smallest number that is a multiple of each of those numbers. For example 12, 24 and 36 are common multiples of 4 and 6, 12 is the lowest common multiple

Getting it right

Write down the first four multiples of 5

5, 10, 15 and 20

Find the lowest common multiple (LCM) of 30 and 42

The first 5 multiples of 42 are
42, 84, 126, 168, 210

Multiples of 30 are
30, 60, 90, 120, 150
180, 210

LCM = 210

WARNING

Don't forget that 5 is a multiple of 5

Start with the bigger number, 42, and keep adding 42
Write down the first 5 multiples to start with.

Keep adding 30 until you come to a number in the first list.

You would get 1 mark just for writing down the two lists of multiples correctly.

Write down the lowest number which appears in both lists.

In Questions **1–5**, write down the first five multiples of each number.

1 3

2 4

3 7

4 11

5 13

In Questions **6** and **7** use the numbers in this list.

5	8	12	17	23	28	36

6 Write down a multiple of 8

7 Write down a multiple of 9

In Questions **8–12**, find the lowest common multiple (LCM) of the two numbers.

8 6, 10   **9** 8, 12   **10** 12, 16   **11** 15, 20   **12** 24, 32

In Questions **13–17**, find the lowest common multiple (LCM) of the three numbers.

13 2, 3, 5

14 2, 3, 4

15 6, 8, 12

16 10, 15, 30

17 6, 9, 12

18 Ben says 'The lowest common multiple of 4 and 6 is 24'.
Is Ben correct?
You must give a reason for your answer.

4 Long multiplication and multiplication of decimals

SKILLS

Use long multiplication

Use long multiplication with decimals

Multiply a decimal by a decimal number

EXAM FACTS

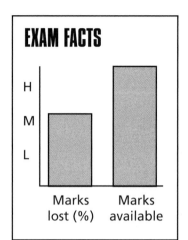

	Marks lost (%)	Marks available

KEY FACTS

- When multiplying decimal numbers the number of decimal places in the answer is the same as the total number of decimal places in the question.

Getting it right

Work out 64×32

Method 1

```
   64
 × 32
 ____
  128
 1920
 ____
 2048
```

Method 2

×	60	4
30		
2		

×	60	4
30	1800	120
2	120	8

×	60	4
30	1800	120
2	120	8
	1920	128

$1920 + 128 = 2048$

$2 \times 64 = 128$

$30 \times 64 = 1920$

WARNING

A common error is to write 192 instead of 1920

This method uses a grid. Split each number into tens and units.

1800 is the answer to 30×60
Complete the grid.

WARNING

$30 \times 60 = 180$ is a common error.

Add the numbers in the columns. Do not include the 60 and the 4

Work out the cost of 23 CDs at £9.75 each.

Cost = 23 × £9.75

```
    975
 ×   23
 ─────
   2925
  19500
 ─────
  22425
```

Cost = £224.25

Work out   0.06 × 0.54

```
   54
 ×  6
 ───
  324
```

0.06 × 0.54 = 0.0324

WARNING

20 × 975 = 1950 is a common error.

Carry out the multiplications ignoring the decimal points.

The number of decimal places in the answer (2) is the same as the total number of decimal places in the question.

EXAM TIP

As a check, 23 × £9.75 is approximately 20 × £10 = £200 so £224.25 is the correct size.

There must be **4** figures after the decimal point in the answer. A decimal point and a zero must be put in front of the 324. Usually, for numbers between 0 and 1, a zero is written before the decimal point.

Now try these

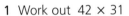

1 Work out 42 × 31

2 Work out 56 × 84

3 Work out 124 × 24

4 Work out 268 × 48

5 'Jet Tours' has an aeroplane that will carry 27 passengers.
Each of the 27 passengers pays £55 to fly from Liverpool to Prague.
Work out the total amount that the passengers pay.

(1387 June 2006)

6 Work out 4.2 × 26

7 Work out 0.56 × 36

8 Nick takes 26 boxes out of his van.
The weight of each box is 32.9 kg.
Work out the total weight of the 26 boxes.

(1387 June 2004)

9 Fatima bought 48 teddy bears at £9.55 each.
Work out the total amount she paid.

(1387 June 2003)

10 Work out
  a 45 × 0.5
  b 0.3 × 0.3
  c 7.2 × 3.6
  d 2.9 × 0.7
  e 0.7 × 0.6
  f 0.03 × 84

SKILLS

Use long division

Use long division with decimals

Divide a decimal by a decimal number

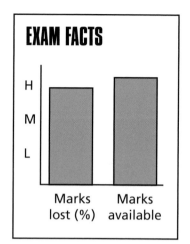

KEY FACTS

- There are many different ways to carry out division, two of which are shown in the first example.

- To divide a decimal number by a whole number use the same method as for long division, remembering to put in the decimal point.

- To divide any number by a decimal number

  - write the division sum as a fraction

  - multiply **both** the numbers by the same power of 10 (10, 100, 1000, etc.) to make the decimal number on the bottom a whole number

  - divide as usual.

Getting it right

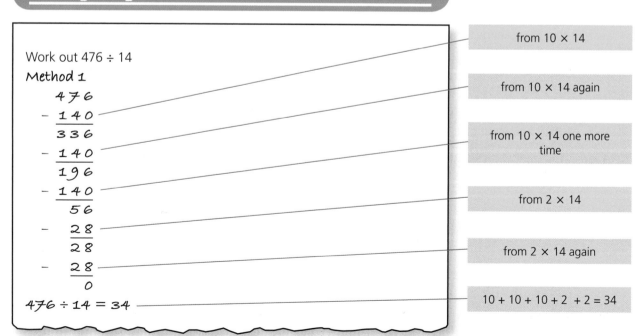

Work out 476 ÷ 14

Method 1

```
      4 7 6
  −   1 4 0
      3 3 6
  −   1 4 0
      1 9 6
  −   1 4 0
        5 6
  −     2 8
        2 8
  −     2 8
          0
476 ÷ 14 = 34
```

from 10 × 14

from 10 × 14 again

from 10 × 14 one more time

from 2 × 14

from 2 × 14 again

10 + 10 + 10 + 2 + 2 = 34

Method 2

14 28 42 56

$$14\overline{)4\ 7^5 6}$$ with quotient 3

$$14\overline{)4\ 7^5 6}$$ with quotient 3 4

$476 \div 14 = 34$

Tim paid £5.44 for 17 pencils.
Each pencil cost the same.
Work out the cost of each pencil.

(1385 November 2000)

17   34   51   68

$$17\overline{)5.^5 4^3 4}$$ with quotient 0. 3

$$17\overline{)5.^5 4^3 4}$$ with quotient 0. 3 2

Each pencil costs £0.32

Eloise bought 16 books for £72
Each book cost the same.
Work out the cost of each book.

16   32   48   64   80

$$16\overline{)7\ 2.^8 0\ 0}$$ with quotient 4. 5 0

Each book costs £4.50

Write down the first few multiples of 14

14 goes into 47 three times with a remainder of 5

14 goes into 56 four times with no remainder

Divide £5.44 by 17 to find the cost of each pencil.

Write down the first few multiples of 17

WARNING

Remember the decimal point in the answer goes above the decimal point in the 5.44

17 goes into 5 zero times with a remainder of 5
17 goes into 54 three times with a remainder of 3

17 goes into 34 two times exactly

The answer may also be written as 32p

Divide £72 by 16 to find the cost of each book.

Write down the first few multiples of 16

Write 72 as 72.00

16 goes into 72 four times with a remainder of 8
16 goes into 80 five times exactly

Don't forget the final zero.

Work out 6 ÷ 0.2

$$6 \div 0.2 = \frac{6}{0.2}$$

$$\frac{6}{0.2}{}^{\times 10}_{\times 10} = \frac{60}{2}$$

$$60 \div 2 = 30$$

$$6 \div 0.2 = 30$$

Write the division sum as a fraction.

To make 0.2 a whole number you must multiply it by 10. So multiply both 6 and 0.2 by 10

Work out 23.1 ÷ 0.03

$$23.1 \div 0.03 = \frac{23.1}{0.03}$$

$$\frac{23.1}{0.03}{}^{\times 100}_{\times 100} = \frac{2310}{3}$$

$$\begin{array}{r} 7\ 7\ 0 \\ 3\overline{)2\ 3\,^21\ 0} \end{array}$$

$$23.1 \div 0.03 = 770$$

To make 0.03 a whole number you must multiply it by 100. So multiply both 23.1 and 0.03 by 100

Then divide in the usual way.

WARNING

A common error is to forget to write the zero and to give the answer as 77

Work out 7.68 ÷ 1.2

$$7.68 \div 1.2 = \frac{7.68}{1.2}$$

$$\frac{7.68}{1.2}{}^{\times 10}_{\times 10} = \frac{76.8}{12}$$

$$\begin{array}{r} 6.\ 4 \\ 12\overline{)7\ 6.\,^48} \end{array}$$

$$7.68 \div 1.2 = 6.4$$

To make 1.2 a whole number you must multiply it by 10. So multiply both 7.68 and 1.2 by 10
The numerator does not have to be a whole number.

Then divide in the usual way.

1 Work out
  a 615 ÷ 15   **b** 552 ÷ 12   **c** 560 ÷ 16   **d** 624 ÷ 24

2 Work out
  a £765 ÷ 15   **b** £340 ÷ 25   **c** £500 ÷ 16   **d** £756 ÷ 24

3 The price of a box of chocolates is £4.32
  There are 24 chocolates in the box.
  Work out the cost of **one** chocolate.

(1385 June 2001)

4 Fatima sold 48 teddy bears for a total of £696
  She sold each teddy bear for the same price.
  Work out the price at which Fatima sold each teddy bear.

(1387 June 2003)

HINT
Write £696 as £696.00

5 Pens cost 25p each.
  Mr Smith spends £120 on pens.
  Work out the number of pens he gets for £120

(1385 November 2001)

6 Mario delivers pizzas.
  He is paid 65p for each pizza he delivers.
  One day he was paid £27.30 for delivering pizzas.
  How many pizzas did Mario deliver?

(1387 November 2004)

HINT
Change £27.30 to pence or change 65p to pounds.

7 Work out 793 ÷ 26

(1385 June 2000)

8 Work out
  a 8 ÷ 0.2   **b** 2.4 ÷ 0.4   **c** 0.36 ÷ 0.9   **d** 8.42 ÷ 0.2
  e 0.32 ÷ 0.02 **f** 4.5 ÷ 0.05   **g** 0.216 ÷ 0.06 **h** 0.732 ÷ 0.03

9 Work out
  a 6.2 ÷ 0.02   **b** 0.936 ÷ 0.03 **c** 1.25 ÷ 0.005 **d** 53 ÷ 0.2
  e 75.2 ÷ 0.4   **f** 6.3 ÷ 0.09   **g** 0.064 ÷ 0.08 **h** 3.24 ÷ 0.6

10 Work out
  a 8.1 ÷ 0.15   **b** 36.8 ÷ 1.6   **c** 0.516 ÷ 0.12 **d** 92.5 ÷ 0.25

SKILL

Find a fraction of a quantity

KEY FACTS

- To find a fraction of a quantity when the numerator is 1, divide the quantity by the denominator of the fraction.
- To find a fraction of a quantity when the numerator is more than 1, first divide the quantity by the denominator and then multiply the result by the numerator.

EXAM FACTS

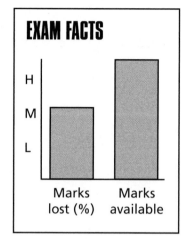

REFERENCE

For a reminder of how to divide, see pages 18 to 20

Getting it right

There are 800 people on a train at Manchester.

$\frac{1}{10}$ of these 800 people are children.

i   Work out $\frac{1}{10}$ of 800

$\frac{3}{8}$ of these 800 people are women.

ii   Work out $\frac{3}{8}$ of 800

(1388 June 2006)

i $\frac{1}{10}$ of 800 = 800 ÷ 10

= 80

ii $\frac{1}{8}$ of 800 = 800 ÷ 8

= 100

$\frac{3}{8}$ of 800 = 3 × 100

= 300

To find $\frac{1}{10}$ of 800, divide 800 by 10

First, find $\frac{1}{8}$ of 800
To find $\frac{1}{8}$ of 800, divide 800 by 8

To find $\frac{3}{8}$ of 800, multiply ($\frac{1}{8}$ of 800) by 3

1 mark would be scored at this stage for a complete correct method.

Danny shares a bag of 20 sweets with his friends.

He gives Mary $\frac{3}{5}$ of the sweets.

He gives Ann $\frac{1}{10}$ of the sweets.

He keeps the rest for himself.

How many sweets does Danny keep for himself?

(1387 June 2006)

$\frac{1}{5}$ of 20 = 20 ÷ 5
= 4

$\frac{3}{5}$ of 20 = 3 × 4
= 12

$\frac{1}{10}$ of 20 = 20 ÷ 10
= 2

12 + 2 = 14

20 – 14 = 6

Danny keeps 6 sweets.

To find $\frac{1}{5}$ of 20,
divide 20 by 5

To find $\frac{3}{5}$ of 20,
multiply ($\frac{1}{5}$ of 20) by 3

To find $\frac{1}{10}$ of 20,
divide 20 by 10

Find the total number of sweets that Danny gives away.

To find the number of sweets Danny keeps, subtract 14 from 20

WARNING

A common error is to forget the final subtraction and give 14 as the answer.

Now try these

1 Work out

   a $\frac{1}{3}$ of 12   b $\frac{1}{4}$ of 40   c $\frac{1}{5}$ of 35   d $\frac{1}{2}$ of 52

   e $\frac{2}{5}$ of 30   f $\frac{3}{10}$ of 120   g $\frac{5}{8}$ of 40   h $\frac{4}{9}$ of 63

2 Work out $\frac{4}{5}$ of 30

  (1388 November 2005)

3 Work out $\frac{2}{5}$ of 20

  (1388 November 2006)

4 Work out $\frac{3}{5}$ of 185

  (1388 January 2003)

5 Work out $\frac{4}{5}$ of 45

6 A hotel has 72 rooms.
$\frac{3}{8}$ of the rooms are empty.
Work out the number of rooms that are **not** empty.

(1385 November 2002)

7 Work out $\frac{3}{7}$ of 168

(1388 November 2005)

8 Work out $\frac{1}{4}$ of £33.56

(1387 June 2005)

9 Work out $\frac{7}{8}$ of £120

(1385 November 2001)

10 In a shop the normal price of a jacket is £60

The cost of the jacket in a sale is $\frac{3}{4}$ of the normal price.

Work out $\frac{3}{4}$ of £60

(1388 March 2004)

11 Judy and Anna share a flat.
The rent for the flat is £125 each week.
Judy pays $\frac{3}{5}$ of the £125
Work out how much rent Judy pays each week.

(1385 November 1999)

12 One day, Alison travelled a total of 145 miles.

She travelled $\frac{2}{5}$ of this distance in the morning.

How many miles did she travel during the rest of the day?

(1387 June 2005)

13 Sophie has 60 DVDs.

$\frac{1}{3}$ of her DVDs are comedies.

$\frac{2}{5}$ of her DVDs are science fiction.

The rest of her DVDs are cartoons.
How many of Sophie's DVDs are cartoons?

14 There are 150 boys in Year 10.

$\frac{1}{5}$ of the boys prefer rugby.

$\frac{3}{10}$ of the boys prefer basketball.

The rest of the boys prefer football.
How many boys prefer football?

Percentages of quantities

SKILL

Find a percentage of a quantity

KEY FACTS

- Per cent means 'out of 100'
- 45% means 45 out of 100 or, as a fraction, $\frac{45}{100}$
- To find a percentage of a quantity
  - change the percentage to a fraction
  - work out that fraction of the quantity.
- You should **learn** these facts
  - $10\% = \frac{1}{10} = 0.1$ $25\% = \frac{1}{4} = 0.25$ $50\% = \frac{1}{2} = 0.5$ $75\% = \frac{3}{4} = 0.75$

EXAM FACTS

| | Marks lost (%) | Marks available |

REFERENCE

For a reminder of how to find fractions of quantities, turn to pages 22 to 23

Getting it right

Work out 20% of 1800   *(1385 June 2001)*

$10\% \text{ of } 1800 = \frac{1}{10} \text{ of } 1800$

$= 1800 \div 10$

$= 180$

$20\% \text{ of } 1800 = 2 \times 180$

$= 360$

> $10\% = \frac{1}{10}$

> To find $\frac{1}{10}$ of a quantity, divide the quantity by 10

> To find 20% of 1800, multiply 10% of 1800 by 2

There are 800 students at Prestfield School.
45% of these 800 students are girls.
Work out 45% of 800   *(1387 June 2004)*

$45\% \text{ of } 800$

$= \frac{45}{100} \times 800$

$= 360$

> $45\% = \frac{45}{100}$
> 'of' means 'multiply'
> Key in 45 ÷ 100 × 800 =

> Writing this working would score 1 mark.

1 Work out
a 50% of 400	b 10% of 60	c 25% of 8
d 75% of 8	e 25% of 120	f 75% of 600

2 Work out
a 20% of £400	b 30% of 50 g	c 5% of 40p
d 40% of 30 kg	e 15% of £200	f 90% of 300 cm

3 Work out 65% of 800

4 Work out 60% of 5300 kg.   *(1385 June 2002)*

5 Work out
a 24% of £80	b 16% of £45	c 73% of 9 m
d 38% of 8 kg	e 45% of £61	f 3% of 70 cm

6 Work out 23% of £64   *(1388 June 2003)*

7 The normal cost of a coat is £94
In a sale the cost of the coat is reduced by 36%.
Work out 36% of £94   *(1388 March 2004)*

8 In France, Fred went to a festival.
There were 650 people at the festival.
16% of the people at the festival were British.
Work out 16% of 650   *(1387 November 2004)*

9 The population of a city is 8 million.
9% of the population are Senior Citizens.
Work out 9% of 8 million.   *(1385 November 2002)*

10 Jo did a maths test.
There were a total of 40 marks for the test.
Jo got 65% of the marks.
Work out 65% of 40   *(1385 June 2000)*

11 There are 1200 students at Howgate School.
55% of the students are girls.
How many of the students are boys?

12 Jenny worked in a bookshop for two weeks.
She is paid £125 per week **plus** 10% of the total value of the books she sells that week.
In the first week, she sold books with a total value of £800
Work out the total amount she was paid in the first week.
                    (1387 November 2005)

HINT
5% is $\frac{1}{2}$ of 10%
So, find 10% of 40p then halve the answer.

HINT
90% is 10% less than 100%
So, find 10% of 300 then subtract this amount from 300

EXAM TIP
When dealing with money, remember that, if the answer is not a whole number of pounds, you must include two digits after the decimal point.

HINT
Notice that the question gives the percentage of girls but asks for the number of boys.

HINT
Work out 10% of £800 and then find the **total** amount she was paid.

8 Percentage change and VAT

SKILLS

Increase or decrease a quantity by a percentage

Calculate VAT

Add on VAT

EXAM FACTS

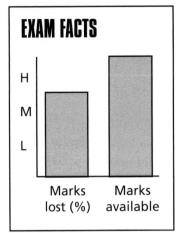

KEY FACTS

- To increase a quantity by a percentage
  - find the given percentage of the original quantity
  - add this amount onto the original quantity.
- To decrease a quantity by a percentage
  - find the given percentage of the original quantity
  - subtract this amount from the original quantity.
- To add on Value Added Tax (VAT) use the same method as a percentage increase.

REFERENCE

For a reminder of how to find a percentage of a quantity, turn to page 25

Getting it right

Martin buys a watch for £24
He sells the watch and makes a profit of 15%.

Work out how much he sells the watch for.

(1388 November 2006)

15% of 24

$= \dfrac{15}{100} \times 24$

$= 3.6$

$24 + 3.6 = 27.6$

Selling price $= £27.60$

> Firstly, work out 15% of 24

> $15\% = \dfrac{15}{100}$
>
> Key in 15 ÷ 100 × 24 =

WARNING ⚠

This is the increase. It is not the final answer as the question asks you to work out how much Martin sells the watch for.

Martin makes a **profit**. This means that he sells the watch for more than he bought it for. Add the profit on to the original price, so add 3.6 onto 24

EXAM TIP

As the answer is a sum of money, there must be two figures after the decimal point or you could miss a mark.

Ann buys a dress in a sale. The normal price of the dress is reduced by 20%. The normal price is £36.80
Work out the sale price of the dress.

(1388 March 2005)

$$20\% \text{ of } 36.80 = \frac{20}{100} \times 36.80$$
$$= 7.36$$
$$36.80 - 7.36 = 29.44$$
Sale price $= £29.44$

The dress is reduced in price which means the price goes down.
So subtract 7.36 from 36.80

Firstly, work out 20% of 36.80

$$20\% = \frac{20}{100}$$

Key in
20 ÷ 100 × 36.80 =

WARNING ⚠

This is the decrease. It is not the final answer as the question asks for the sale price of the dress.

Mr Brown buys a garden spade.
The spade costs £20 plus $17\frac{1}{2}\%$ VAT.
Calculate the total cost of the spade.

(1387 November 2006)

$$10\% \text{ of } 20 = \frac{1}{10} \text{ of } 20$$
$$= 20 \div 10$$
$$= 2$$

$$5\% \text{ of } 20 = \frac{1}{2} \times 10\% \text{ of } 20$$
$$= \frac{1}{2} \times 2$$
$$= 1$$

$$2\frac{1}{2}\% \text{ of } 20 = \frac{1}{2} \times 5\% \text{ of } 20$$
$$= \frac{1}{2} \times 1$$
$$= 0.5$$

$$17\frac{1}{2}\% \text{ of } 20 = 2 + 1 + 0.5$$
$$= 3.5$$

$$20 + 3.5 = 23.5$$
Total cost of spade $= £23.50$

To find 5% of 20, find $\frac{1}{2}$ of (10% of 20).

VAT is a tax so the cost of the spade **increases**, so add 3.5 on to 20

$$17\frac{1}{2}\% = 10\% + 5\% + 2\frac{1}{2}\%$$

Firstly, work out 10% of 20
$$10\% = \frac{1}{10}$$

To find $2\frac{1}{2}$% of 20, find $\frac{1}{2}$ of (5% of 20)

WARNING ⚠

This is the VAT. It is not the final answer as the question asks you to work out the total cost.

EXAM TIP

As the answer is a sum of money, there must be two figures after the decimal point.

Now try these

1 Emily bought a car for £1400
She sold the car.
She made a profit of 15%.
Work out how much she sold the car for.

2 The price of a DVD player was £180
In a sale, the price is reduced by 20%.
Work out the sale price of the DVD player.

3 Jane is going to buy a computer for £480 + $17\frac{1}{2}$% VAT.
Work out the total price, including VAT, that Jane will pay for the computer.
(1385 November 2000)

HINT

Emily makes a profit so the price **increases**.

HINT

The DVD player is reduced so the price **decreases**.

4 A year ago, Jack weighed 80 kg.
Jack now weighs 5% less.
Work out how much Jack now weighs.

5 A compact disc player costs £50 plus $17\frac{1}{2}\%$ VAT.

Calculate the total cost of the compact disc player.

(1387 November 2005)

6 The price of a jacket is £80
In a sale, the price is reduced by 15%.
Work out the sale price of the jacket.

7 Tom is going to buy a computer game, the usual price of which is £40
The usual price of the computer game is reduced by 18%.
Work out the price Tom pays.

8 Hayley's salary is £34 000
Her salary increases by 6%.
Work out Hayley's new salary.

9 Wayne bought an engagement ring for Tracy.
The total cost of the ring was £420 **plus** VAT at $17\frac{1}{2}\%$.
Work out the cost of the ring.   *(1387 November 2003)*

10 In a sale, normal prices are reduced by 12%.
The normal price of a camera is £86
Work out the sale price of the camera.

11 The price of all rail season tickets to London increases by 4%.
Before this increase, the price of a rail season ticket from Reading to
London was £2664
Work out the price after the increase.   *(1387 November 2006)*

12 Daryl buys a car for £9800
The value of the car depreciates by 35% each year.
Work out the value of the car at the end of one year.

HINT

"Depreciates" means that the
value of the car decreases.

13 Mrs Brown bought 32 pens for a total of £11.20
She sold all of the pens.
She made a profit of 20% on each pen.
Work out how much she sold each pen for.   *(1388 March 2002)*

HINT

First work out how much
Mrs Brown paid for **each** pen.

14 Martin had to buy some cleaning materials.
The cost of the cleaning materials was £64.00 plus VAT at $17\frac{1}{2}\%$.
Work out the total cost of the cleaning materials.

(1387 June 2004)

15 Kunal sells CDs.
He sells each CD for £8.40 plus VAT at $17\frac{1}{2}\%$.
He sells 1200 CDs.
Work out the total amount Kunal receives.

16 The total cost of orange drink for 50 cups is £7.50
Each cup of drink is sold at a 20% profit.
Work out the price at which each cup of drink is sold.

(1385 November 2000)

9 Using a calculator

SKILL

Use a calculator efficiently

KEY FACTS

- The order in which calculations are carried out is given by BIDMAS

  Brackets
  Indices
  Division
  Multiplication
  Addition
  Subtraction

- One number written over another as a fraction, is worked out by division. For example $\frac{10.8}{7.2}$ means the same as $10.8 \div 7.2$

EXAM FACTS

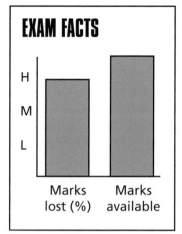

Getting it right

Use a calculator to work out

4.5×3.2^2

Write down all the figures on your calculator display.

4.5×10.24

$= 46.08$

$4.5 \times 3.2^2 = 46.08$

Work out 3.2^2 (Indices) first and write down the answer.

Multiply to complete the calculation.

If you have a scientific calculator, key in
$\boxed{4.5}\,\boxed{\times}\,\boxed{3.2}\,\boxed{x^2}\,\boxed{=}$

Note
The software in scientific calculators is designed so that you can key in 'what you see'. You will also get the correct result if you follow the rules of BIDMAS:
$\boxed{3.2}\,\boxed{x^2}\,\boxed{\times}\,\boxed{4.5}\,\boxed{=}$

EXAM TIP

Check by doing an estimate:
$5 \times 3^2 = 5 \times 9 = 45$

Use a calculator to work out

$$\frac{\sqrt{3.24 + 2.96}}{8.645 - 2.81}$$

Write down all the figures on your calculator display.

$$\frac{\sqrt{3.24 + 2.96}}{8.645 - 2.81}$$

$$= \frac{\sqrt{6.2}}{5.835}$$

$$= \frac{2.48997992}{5.835}$$

$$= 0.426731777$$

WARNING

In the expression $\sqrt{3.24 + 2.96}$ the addition has to be done first, before the square root.
A common error is to work out just $\sqrt{3.24}$ getting $1.8 + 2.96 = 4.76$

Work out the value of the denominator (the bottom line) and write it down.

Work out the value of the numerator (the top line) and write down the full calculator display.

Divide the numerator by the denominator and write down the full calculator display.
Different models of calculator will give slightly different answers.

Now try these

Work out each of the following.
Write down all the figures on your calculator display.

1 $2.52 + 2.5^2$ 2 $\sqrt{2} \times 3.58$ 3 $\dfrac{6.42 - 3.48}{5.66}$

4 $8.77 - \dfrac{16.2}{2.25}$ 5 $\dfrac{4.7 + 2.6^2}{1.25}$ 6 $\dfrac{(6.2 - 3.9)^2}{1.25}$

(1387 Mock 2002)

7 $\sqrt{2.1^2 + 2.8^2}$ 8 $\dfrac{\sqrt{9.3} - 2.4}{\sqrt{9.3} + 2.4}$

9 $(2.3 + 1.8)^2 \times 1.07$ *(1387 June 2003)*

10 $\sqrt{46} - 2.5^2$ *(1388 June 2003)*

EXAM TIP

For the above question, if you use a scientific calculator, then you should follow the same method, as there may be marks for showing the values at each stage.

SKILLS

Share a quantity in the ratio $a : b$

Share a quantity in the ratio $a : b : c$

EXAM FACTS

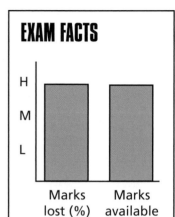

KEY FACTS

- To share a quantity in a given ratio
  - find the total number of shares by adding the numbers in the ratio
  - work out what each share is worth by dividing the quantity by the total number of shares
  - work out the answers by multiplying the value of each share by each of the numbers in the ratio.

Getting it right

Graham and Keith share £300 in the ratio 2 : 3
Work out how much each receives. *(1388 November 2006)*

$2 + 3 = 5$
$300 \div 5 = 60$
$60 \times 2 = 120$
$60 \times 3 = 180$

Graham receives £120
Keith receives £180

Add 2 and 3 to get the total number of shares.

Divide 300 by 5 to work out what each share is worth. This would score 1 mark.

Graham gets 2 shares so multiply 60 by 2

Keith gets 3 shares so multiply 60 by 3

Check that the sum of the amounts is 300
(120 + 180 = 300)

WARNING

Check that each amount of money goes with the correct person.

Three women earned a total of £36
They shared the £36 in the ratio 7 : 3 : 2
Donna received the largest amount.
Work out the amount Donna received. *(1387 June 2005)*

$7 + 3 + 2 = 12$
$36 \div 12 = 3$
$3 \times 7 = 21$

Donna received £21

Add 7, 3 and 2 to get the total number of shares.

Divide 36 by 12 to work out what each share is worth.

Donna receives the largest share.
7 is the largest number in the ratio 7 : 3 : 2
So Donna receives 7 shares.

Donna gets 7 shares so multiply 3 by 7

Now try these

1 a   Share £45 in the ratio 2 : 3
   b   Share £36 in the ratio 4 : 5
   c   Share £140 in the ratio 1 : 6
   d   Share £100 in the ratio 2 : 3 : 5
   e   Share £350 in the ratio 1 : 2 : 4

2   Kamini and David share £84 in the ratio 1 : 3
   Calculate the amount of money Kamini gets.

(1388 June 2006)

3   Judy and Anna share a flat.
   They hire a television.
   The hire cost is £24.50 each month.
   Judy and Anna share the £24.50 in the ratio 3 : 4
   Work out how much each of them pays for the television each month.

(1385 November 1999)

4   Jenny and Kath hire the canal boat for 14 days.
   They share the hire cost of £1785.00 in the ratio 2 : 3
   Work out the smaller share.

(1387 November 2005)

> **Canal boat for hire**
> **£1785.00**
> **for 14 days**

5   Verity and Jean share £126 in the ratio 5 : 3
   Work out how much money Verity receives.

(1388 March 2003)

6   On a school trip, the ratio of the number of teachers to the number of
   students was 1 : 7
   The total number of teachers and students was **160**
   Work out the number of teachers on this school trip.

(1388 November 2005)

7   Ahmed, Ben and Cara were given a total of £720
   They shared the money in the ratio 3 : 2 : 7
   Work out how much money each person received.

(1388 January 2002)

8   Mrs Jones shared £357 between her two children in the ratio 1 : 6
   She gave the bigger share to Matthew.
   Work out how much money she gave to Matthew.

(1388 March 2004)

9   Amy, Beth and Colin share 36 sweets in the ratio 2 : 3 : 4
   Work out the number of sweets that each of them receives.

(1387 June 2006)

10   Mr Brown makes some compost.
   He mixes soil, manure and leaf mould in the ratio 3 : 1 : 1
   Mr Brown makes 75 litres of compost.
   How many litres of soil does he use?

(1387 November 2006)

Currency exchange rates

SKILL

Use exchange rates to work out equivalent amounts in different currencies

EXAM FACTS

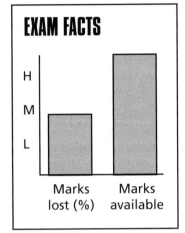

KEY FACTS

- To change pounds into another currency, *multiply* by the amount that each pound is worth
- To change another currency into pounds, *divide* by the amount that each pound is worth

Getting it right

Sarah went to Germany.
She changed £300 into euros.
The exchange rate was £1 = 1.64 euros.

a   Work out the number of euros Sarah got.

Sarah came home.
She had 119 euros left.
The new exchange rate was £1 = 1.50 euros.

b   Work out how much Sarah got in pounds for 119 euros.

(1388 January 2003)

a $300 \times 1.64 = 492$

Sarah got 492 euros

b $119 \div 1.50 = 79.333...$
$= 79.33$

Sarah got £79.33

Notice that the exchange rate changes for the second part.

Sarah got 1.64 euros for every £1, so *multiply* the number of pounds by 1.64

Sarah got £1 for every 1.50 euros so *divide* the number of euros by 1.50

WARNING

A common error is to multiply in both parts.

EXAM TIP

Write down at least three decimal places of the calculator display.

WARNING

The question asks for the answer in pounds. This does **not** mean round your answer to the nearest pound. The amount should be rounded to the nearest penny (two figures after the decimal point).

Sangita is on holiday in Switzerland.
She buys a train ticket.
She can pay either 100 Swiss francs or
70 euros.

  £1 = 2.10 Swiss francs
  £1 = 1.40 euros

She pays in Swiss francs rather than euros.
Work out how much she saves.
Give your answer in pounds.
(1387 November 2005)

$100 \div 2.10 = 47.61904...$
$ = £47.62$
$70 \div 1.40 = 50$
$ = £50$
$50 - 47.62 = 2.38$

Sangita saves £2.38

The exchange rate for £1 is given for each currency. So the easiest way to answer this question is to convert both prices into pounds.

Every £1 is worth 2.10 Swiss francs so *divide* the number of Swiss francs by 2.10

(1388 November 2005)

WARNING

The most common error that occurred in this question was to multiply rather than divide each currency by the exchange rate.

EXAM TIP

Write down at least three decimal places of the calculator display.

As the answer is a sum of money, give the answer correct to the nearest penny by rounding to two decimal places.

The ticket costs the equivalent of £47.62 if paid for in Swiss francs.

Every £1 is worth 1.40 euros so *divide* the number of euros by 1.40

The ticket costs the equivalent of £50 if paid for in euros.

To work out how much Sangita saves, subtract £47.62 from £50

WARNING

The question asks for the answer in pounds. This does **not** mean round your answer to the nearest pound. The amount should be rounded to the nearest penny (two figures after the decimal point).

Now try these

1. Pete went to South Africa.
  He changed £200 into Rand.
  The exchange rate was £1 = 11.38 Rand.
  Work out the number of Rand that Pete got.

2. Maria went to Rome.
  She changed £350 into euros.
  The exchange rate was £1 = 1.60 euros
  Work out the number of euros that Maria got.

3. Matt returns from a holiday in America.
  He changes $150 back into pounds.
  The exchange rate is £1 = $1.85
  How much money should he get?
  Give your answer to the nearest penny.

4. Nasmeen returns from a holiday in Germany.
  She changes 120 euros into pounds.
  The exchange rate is £1 = 1.62 euros.
  Work out how much money she should get.
  Give your answer to the nearest penny.

5 Karl went to America.
  He changed £180 into dollars.
  The exchange rate was £1 = $1.87
  Work out the number of dollars that Karl got.

6 a Sunita went to Austria.
     She changed £250 into euros.
     The exchange rate was £1 = 1.58 euros.
     Work out the number of euros Sunita got.
  b Sunita came home. She had 80 euros left.
     The new exchange rate was £1 = 1.52 euros.
     Work out how much Sunita got, in pounds, for 80 euros.

7 Bill goes on holiday to New York.
  The exchange rate is £1 = 1.545 dollars.
  He changes £800 into dollars.
  a How many dollars should he get?
  After his holiday, Bill changes 120 dollars back into pounds.
  The exchange rate is the same.
  b How much money should he get?
     Give your answer to the nearest penny.   *(1385 June 2002)*

8 Kylie went to Geneva.
  She changed £200 into Swiss francs.
  The exchange rate was £1 = 2.43 Swiss francs.
  a Work out the number of Swiss francs Kylie got.
  Kylie brought 75 Swiss francs back from Geneva.
  The exchange rate was now £1 = 2.50 Swiss francs.
  b Work out how much Kylie got in pounds.

9 Will buys an MP3 player in England for £120
  He then goes on holiday to America and sees an identical MP3 player for $189
  The exchange rate is £1 = 1.80 dollars.
  In which country is the MP3 player cheaper and by how much?

10 A British family are on holiday in San Francisco.
   At a café they order 3 hot dogs and 1 chicken salad.
   The exchange rate is £1 = $1.44
   Work out their **total** bill in pounds (£).   *(1388 March 2004)*

Menu	
Hot dog	$5.10
Chicken salad	$4.50
Hamburger	$3.80
Pizza	$4.00

11 Mr and Mrs Smith and their 3 children went on holiday to Paris.
   They visited a museum.
   The tickets cost 9 euros for each adult and 5 euros for each child.
   The exchange rate was £1 = 1.62 euros.
   Work out the total cost of the tickets to the museum.
   Give the total cost in pounds.

12 Hugh went on holiday to Italy.
   While on holiday, he went shopping.
   He bought a belt and a hat.
   The belt cost 25 euros.
   The hat cost 14 euros.
   The exchange rate was £1 = 1.56 euros.
   Work out the total cost of the belt and the hat.
   Give the **total** cost in pounds.   *(1388 March 2005)*

SKILL

Estimate a solution to a calculation by rounding numbers correct to 1 significant figure

EXAM FACTS

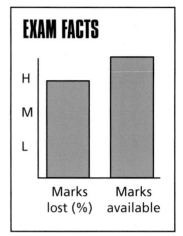

KEY FACT

- A number rounded correct to 1 significant figure (1 s.f.) has only one figure which is not a zero. For example, 239 = 200 correct to 1 significant figure, as 2 is the most significant figure (highest value digit). Also, 0.047 = 0.05 correct to 1 significant figure, as the most significant figure is the 4 and the digit after the 4 is more than 5

REFERENCE

For a reminder of how to multiply decimals, turn to page 17
For a reminder of how to divide by decimals, turn to pages 19 to 20

Getting it right

WARNING

To estimate, first write each number correct to 1 s.f. A common error is to write 640 instead of 600, that is round to 2 s.f. instead of 1 s.f.

Work out an estimate for the value of

$$\frac{637}{3.2 \times 9.8}$$

$$\frac{600}{3 \times 10}$$

Work out the values of the top and bottom lines separately.

$$= \frac{600}{30}$$

$$= 20$$

Work out an estimate for

$$\frac{412 \times 5.904}{0.195}$$

$$\frac{400 \times 6}{0.2}$$

$$= \frac{2400}{0.2}$$

$$= \frac{24\,000}{2}$$

$$= 12\,000$$

To make the division easier, multiply both the top and bottom lines by 10

Now try these

In Questions **1–8**, work out an estimate for the calculation.

1 21.3×29.8

2 $\dfrac{186}{39.9}$

3 $\dfrac{12.1 \times 98}{5.1}$

4 $\dfrac{301}{6.1 \times 5.1}$

5 0.48×3.22

6 $\dfrac{215 \times 0.55}{2.2}$

7 $\dfrac{813 \times 19.8}{97.6}$

(1388 June 2003)

8 $\dfrac{5.79 \times 312}{0.523}$

(1387 November 2005)

9 On average, Nick walks 18 000 steps every day.
He walks 1 mile approximately every 3500 steps.
Work out an estimate for the average distance, in miles,
that Nick walks **in one year**.

(1387 November 2004)

10 The area of a field is 4188 m². Fertiliser is spread on the field at a rate of
8 grams per m². The fertiliser costs £1.95 per kilogram. Work out an
estimate for the total cost of the fertiliser.

SKILLS

Simplify algebraic terms
Simplify by collecting like terms
Solve linear equations with letters on both sides

EXAM FACTS

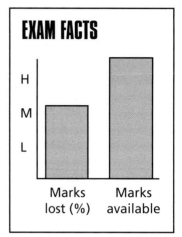

KEY FACTS

- To add a letter n times, multiply the letter by the number n.

- To simplify expressions with numbers, different letters and multiplication signs, remove the multiplication signs.

- To collect like terms, combine the terms which contain the same letter.

- To solve simple linear equations collect the letter terms on one side of the equals sign and the number terms on the other side.

Getting it right

Simplify $5p \times 2q$

$$5p \times 2q = (5 \times 2) \times (p \times q)$$
$$= 10 \times pq$$
$$= 10\,pq$$

Multiply the numbers together.

Multiply the letters together.

WARNING

Common errors are to write the answer as $5p2q$ or $7pq$.

Simplify $2xy + 4xy - xy$

$$2xy + 4xy - xy = 6xy - xy$$
$$= 5xy$$

$2xy + 4xy = 6xy$

xy is the same as $1xy$

WARNING

Leaving the answer as $6xy - xy$ would not get full marks.

Simplify $3c - 5d - c + 2d + 6$

$3c - 5d - c + 2d + 6$

$= (3c - c)(-5d + 2d) + 6$

$= 2c - 3d + 6$

> Combine the terms which contain the same letter.

> $3c - c = 2c$ and
> $-5d + 2d = -3d$
> writing either $2c$ or $-3d$
> would get you 1 mark.

Solve $6x - 5 = 2x + 9$ *(1387 November 2006)*

$6x - 5 = 2x + 9$

$6x - 5 - 2x = 2x + 9 - 2x$

$4x - 5 = 9$

$4x - 5 + 5 = 9 + 5$

$4x = 14$

$x = \dfrac{14}{4} = 3.5$

> Subtracting $2x$ from both sides gets 1 mark. Then collect like terms.

> Adding 5 to both sides gets 1 mark.

> Divide both sides by 4

Now try these

1 Simplify **a** $m + m + m + m + m + m$ **b** $n^2 + n^2$ **c** $pq + pq + pq$

2 Simplify **a** $a \times b \times c$ **b** $2 \times d \times e$ **c** $2x \times 9y$

3 Simplify **a** $8xy + 3xy - 2xy$ **b** $3mn - 5mn + mn$

4 Simplify **a** $e + f + e + f + e$ **b** $3g + 2h + 5h + g$

5 Simplify $4p - 5q + 3p - 3q$

6 Simplify $3a + 5b - a + 2b + 8$ *(1387 July 2006)*

7 **a** Solve the equation $4x + 3 = 19$ *(1387 July 2006)*
   b Solve the equation $4y + 3 = y + 19$

8 Solve the equation $5x - 4 = 3x + 18$

9 Solve $6 + 7x = 3x + 4$

10 Solve $8y + 1 = 3y - 6$

SKILLS

Derive an expression with at least two terms

Derive a formula with at least two unknowns

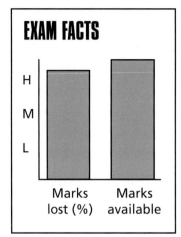
KEY FACTS

- To write an algebraic expression use letters to represent quantities. For example, $2b + 3t$ is an expression with two terms which could represent the total number of wheels on b bicycles and t tricycles.

- A formula is used to describe a rule or a relationship; it must have an equals sign. For example, $W = 2b + 3t$ is a formula where W is the total number of wheels.

REFERENCE

For a reminder of how to simplify expressions by collecting like terms turn to pages 39 to 40

Getting it right

Mr Smith owns minibuses and coaches.
Each minibus has 12 seats. Each coach has 48 seats.
Write an expression, in terms of m and c, for the total number of seats in m minibuses and c coaches.

Number of seats in m minibuses = 12 × m
$\qquad\qquad\qquad\qquad\qquad\qquad = 12m$

Number of seats in c coaches = 48 × c
$\qquad\qquad\qquad\qquad\qquad = 48c$

Total number of seats = 12m + 48c

Writing either of the terms $12m$ or $48c$ would get you 1 mark.

To find the **total** number of seats add the two terms together. This would get the final mark.

WARNING

$12m + 48c$ cannot be simplified so an answer such as $60mc$ would lose the final mark.

The height of a hedge is now 80 cm.
The hedge grows 70 cm higher every year.
Write down a formula for the height, h cm, of the hedge t years from now.
(1387 November 2006)

After 1 year the hedge grows 1 × 70 cm higher.
After 2 years the hedge grows 2 × 70 cm higher.

After 3 years the hedge grows 3 × 70 cm higher.
After t years the hedge grows t × 70 = 70t cm higher.

Height of hedge after t years = 80 + 70t
$$h = 80 + 70t$$

Writing 70t would get 1 mark.

We know that the hedge is already 80 cm high so after t years it will be 80 + 70t.

Writing 80 + 70t would get 1 mark.

Writing a formula for the height, h cm, means starting with h =

Now try these

1 Shirts cost £S each and blouses cost £b each. Write down an expression in terms of S and b for the total cost of 8 shirts and 5 blouses.

2 A box contains 60 chocolates. Trevor eats m milk chocolates and p plain chocolates. Write down an expression, in terms of m and p, for the number of chocolates left in the box.

3 Olivia buys p packs of nappies. Each pack contains n nappies.
   a Write down an expression, in terms of p and n, for the total number of nappies she buys.
   b Olivia pays £25 for these nappies. Write down an expression, in terms of p and n, for the cost of one nappy in pounds.

4 Brody earns £P per hour. Last week he worked for t hours and also earned a bonus of £23. Write down a formula for the total amount, £A, he earned last week.

5 The three angles, in degrees, of a triangle are x, y +1 and 2y.
   Write down a formula for x in terms of y.
   Give your answer in its simplest form.

HINT
The angle sum of a triangle is 180°.

6 In a game, 5 points are awarded for a win, 3 points are awarded for a draw and 1 point is deducted for a loss.
   In a competition, Tyler wins p games, draws q games and loses r games.
   T is the total number of points Tyler is awarded. Write down a formula for T in terms of p and q and r.

7 The diagram shows a trapezium.
   All the lengths are in centimetres.
   The perimeter of the trapezium is P cm.
   Find a formula, in terms of x and y for P.
   Give your answer in its simplest form.   (1388 March 2006)

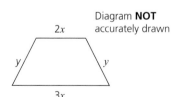
Diagram **NOT** accurately drawn

8 Mrs Hughes has b boxes of crayons. There are c crayons in each box. She shares the crayons amongst the p children in her class. Each child gets x crayons and there are none left over. Write down a formula for x in terms of b, c and p.

9 Gil buys r roses at 60p each and c carnations at 47p each.
   Gil pays with a £50 note.
   He should get £H change.
   Write down a formula for H in terms of r and c.

15 Multiplying out brackets

SKILLS

Multiply out brackets

Multiply out brackets and simplify

Expand and simplify $(x + a)(x + b)$

EXAM FACTS

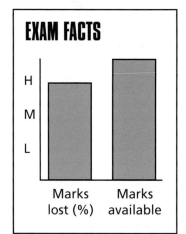

| Marks lost (%) | Marks available |

KEY FACTS

- To multiply out brackets, multiply each term inside the bracket by the term outside the bracket.
- To multiply out brackets and simplify, multiply out the brackets and collect like terms.
- To expand brackets of the form $(x + a)(x + b)$, multiply each term in the first bracket by each term in the second bracket and then collect like terms.

REFERENCE

For a reminder of how to simplify expressions by collecting like terms turn to pages 39 to 40

Getting it right

Multiply y by y and multiply -5 by y

Expand $y(y - 5)$ *(1387 November 2006)*

$$y(y - 5) = y \times y - 5 \times y$$
$$= y^2 - 5y$$

Remember $y \times y = y^2$

WARNING

A common error is to multiply only the first term in the bracket by y.
$y^2 - 5$ gets no marks.

Expand and simplify $2(3m + 4) + 3(m - 5)$

                (1387 November 2006)

$$2(3m + 4) + 3(m - 5)$$
$$= 2 \times 3m + 2 \times 4 + 3 \times m + 3 \times -5$$
$$= 6m + 8 + 3m - 15$$
$$= 9m - 7$$

Multiply out each bracket for 1 mark.

positive × negative = negative

Collect like terms for final mark.

Expand and simplify $(x - 2)(x - 3)$

$$(x - 2)(x - 3) = x(x - 3) - 2(x - 3)$$
$$= x^2 - 3x - 2x + 6$$
$$= x^2 - 5x + 6$$

Multiply the second bracket, $(x - 3)$, by each term in the first bracket.

negative × negative = positive

The expansion of these brackets always leads to 4 terms. Writing these 4 terms correctly gets 1 mark.

Now try these

In Questions **1–4**, multiply out.

1 a $3(x + 1)$ **b** $5(2 - y)$

2 a $a(b + 4)$ **b** $c(d - e)$

3 a $10(2p - 3)$ **b** $7(7 - 4q)$

4 a $2x(5y - 1)$ **b** $3a(1 + 3b)$

5 Expand   **a** $x(x - 1)$ **b** $y(3 - 2y)$

In Questions **6–20**, expand and simplify.

6 $4 + 3(2a + 1)$

7 $8 - 5(2b + 1)$

8 $3(a + b) + 2(3a + 2b)$

9 $2(c - 2d) + 3(2c + d)$

10 $2(m - 2) + 3(3m - 4)$

11 $4(x - 3y) - 2(3x + y)$

12 $5(2z - 3) - 2(3 - z)$

13 $(x + 5)(x + 3)$ *(1387 November 2006)*

14 $(x + 2)(x - 1)$ **15** $(y - 3)(y + 7)$

16 $(z - 8)(z + 3)$ **17** $(p - 4)(p - 5)$

18 $(2 + x)(3 + x)$ **19** $(y + 1)^2$

20 $(x - 4)^2$

16 Negative numbers, expressions and formulae

SKILLS

Add and subtract positive and negative numbers

Multiply and divide using positive and negative numbers

Substitute negative numbers into expressions and formulae

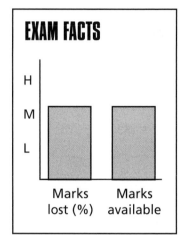
KEY FACTS

- To find the value of an expression, substitute the value of each letter and work out the result.

- To find the value of the subject of a formula, substitute given values for the letters on the right-hand side of the = sign.

Getting it right

In the expression, replace each letter by its value – this gets 1 mark.

Work out the value of $p + 3q$
when $p = 2$ and $q = -1.5$

$$p + 3q = 2 + 3 \times (-1.5)$$

$$= 2 - 4.5$$

$$= -2.5$$

WARNING

Multiply before Add
(BIDMAS)
A common error is to add the 2 and 3 first and write
$5 \times (-1.5)$

positive $\times$ negative = negative
negative $\times$ positive = negative
positive $\times$ positive = positive
negative $\times$ negative = positive

$P = 2x^2 + 3$
Find the value of P when $x = -5$

(1388 June 2006)

$P = 2 \times (-5)^2 + 3$

$P = 2 \times (+25) + 3$

$P = 50 + 3$

$P = 53$

In the formula, replace the letter on the right-hand side by its value – this gets 1 mark.

WARNING

$2 \times (-5)^2$ is 2 times $(-5)^2$
A common error is to multiply (-5) by 2 and work out $(-10)^2$

WARNING

A common error is to write $(-5)^2 = -25$
Remember
negative $\times$ negative = positive
so
$(-5)^2 = (-5) \times (-5) = +25$

Now try these

1 Work out the value of $a + bc$ when $a = -3$, $b = 2$ and $c = 5$

2 Work out the value of $pq + 3r$ when $p = -5$, $q = 4$ and $r = 6$

3 Find the value of $\dfrac{x + y}{2}$ when $x = -5$ and $y = -7$

4 Find the value of $x^2 - 5x$ when **a** $x = 3$ **b** $x = 0$ **c** $x = -4$

5 Uzma says that, when $n = -5$, the value of $3n^2$ is 75. Atif says that the correct answer is 225 and Simon says it is -225. Who is right? Explain your answer.

6 Find the value of $(x - 4)^2$ when **a** $x = 1$ **b** $x = -6$

7 $S = 2p + 3q$

$p = -4$, $q = 5$

Work out the value of S. *(1387 June 2006)*

8 $P = a^2 + 5b$

$a = -3$, $b = 2$

Work out the value of P. *(1388 November 2006)*

9 The formula $C = \dfrac{5(F - 32)}{9}$ can be used to change temperatures in degrees Fahrenheit (F) to temperatures in degrees Celsius (C).
a Change 14 degrees Fahrenheit to degrees Celsius.
b Change -58 degrees Fahrenheit to degrees Celsius.

10 The formula $s = ut + \frac{1}{2}at^2$ is used in science.
Work out the value of s when $u = -40$, $t = 4$ and $a = -10$

17 Tables of values

SKILL

Substitute values into an equation to complete a table of values for a graph

KEY FACTS

- To complete a table of values, substitute the given values of x into an equation of the form $y = ax + b$ or $y = ax^2 + bx + c$ to find the unknown values of y.

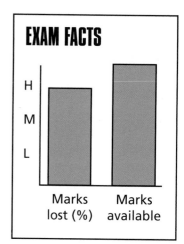
REFERENCE

For a reminder of how to substitute negative numbers into expressions turn to pages 45 to 46

Getting it right

Complete the table of values for $y = 3 - 2x$

Multiply before Subtracting (BIDMAS)

x	−3	−2	−1	0	1	2	3
y		7			1		−3

When $x = 2$, $y = 3 - 2 \times 2 = 3 - 4 = -1$

When $x = 0$, $y = 3 - 2 \times 0 = 3 - 0 = 3$

When $x = -1$, $y = 3 - 2 \times (-1) = 3 + 2 = 5$

When $x = -3$, $y = 3 - 2 \times (-3) = 3 + 6 = 9$

x	−3	−2	−1	0	1	2	3
y	9	7	5	3	1	−1	−3

Show your substitution of the values of x. There will usually be space to do this. Showing your working out reduces the number of errors.

WARNING

A common error is to write $2 \times 0 = 2$

negative × negative = positive

Two correct values for y for 1 mark.

Notice the pattern. The numbers 9, 7, 5, ... decrease by 2 each time. There will always be a pattern for equations of the form $y = ax + b$.

Complete the table of values for
$y = x^2 - 3x + 1$

x	−2	−1	0	1	2	3	4
y	11		1	−1			5

(1387 June 2006)

When $x = 3$, $y = 3^2 - 3 \times 3 + 1$
$= 9 - 9 + 1$
$y = 1$

When $x = 2$, $y = 2^2 - 3 \times 2 + 1$
$= 4 - 6 + 1$
$y = -1$

When $x = -1$, $y = (-1)^2 - 3 \times (-1) + 1$
$= 1 + 3 + 1$
$y = 5$

x	−2	−1	0	1	2	3	4
y	11	5	1	−1	−1	1	5

WARNING

A common error is to write
$(-1)^2 = -1$

One or two correct values
for y for 1 mark.

Notice the symmetry

$$5 \quad 1 \quad -1 \mid -1 \quad 1 \quad 5$$

This is an indication
that your values for y are
correct.

Look for symmetry when
the equation is of the form
$y = ax^2 + bx + c$

Now try these

1 Complete the table of values for
$y = x + 8$

x	−3	−2	−1	0	1	2	3
y	5		7				11

2 Complete the table of values for
$y = 9 - 5x$

x	−2	−1	0	1	2	3	4
y		14		4			−11

3 Complete the table of values for
$y = x^2 + 6$

x	−3	−2	−1	0	1	2	3
y	15			6		10	

4 Complete the table of values for
$y = 2x^2 - 7x + 4$

x	−4	−3	−2	−1	0	1	2
y	64		26			−1	−2

5 Complete the table of values for
$y = 7 + 4x - 4x^2$

x	−2	−1	0	1	2	3	4
y			7		−1		−41

Graphs

18

SKILLS

Use a table of values to draw a simple straight line graph

Draw the graph of a quadratic function from a table of values

EXAM FACTS

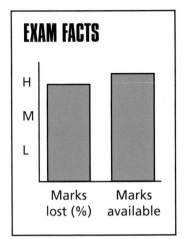

Marks lost (%) Marks available

KEY FACTS

- To draw a simple straight line graph, plot the points from the table of values and join them with one straight line.
- To draw the graph of a quadratic function, plot the points from the table of values and join them with a smooth curve.

REFERENCE

For a reminder of how to construct and complete a table of values turn to pages 47 to 48

Getting it right

This table of values is for $y = 3 - 2x$

x	−3	−2	−1	0	1	2	3
y	9	7	5	3	1	−1	−3

a   On the grid, draw the graph of $y = 3 - 2x$

b   Use your graph to find

   i   the value of y when $x = -1.7$

   ii the value of x when $y = -1.6$

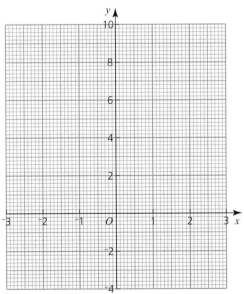

a

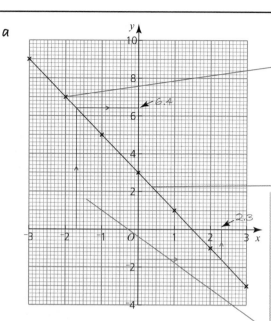

Plot each of the 7 points in the table with a cross (×) to get 1 mark.

WARNING

Notice the scales on the *x*-axis and *y*-axis are different.

Join the 7 points with **one** straight line to get 1 mark.

WARNING

WARNING

Take care with the scale when reading off the values on the axes.
A common wrong answer for **b i** is $y = 6.2$

A line which does not go from $(-3,9)$ to $(3, -3)$, for example a line which stops at $(-2,7)$ would not get full marks.

If one of your plotted points does not lie on the line check your table of values.

b **i** When $x = -1.7$, $y = 6.4$
ii When $y = -1.6$, $x = 2.3$

The red and green lines show the methods.

This table of values is for $y = x^2 - 3x + 1$

x	−2	−1	0	1	2	3	4
y	11	5	1	−1	−1	1	5

a On the grid, draw the graph of $y = x^2 - 3x + 1$

b Use your graph to estimate the values of x for which $y = 3$

(1387 June 2006)

"value**s**" means that there is more than one value of x

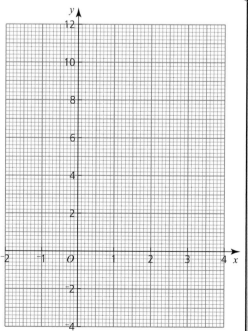

a

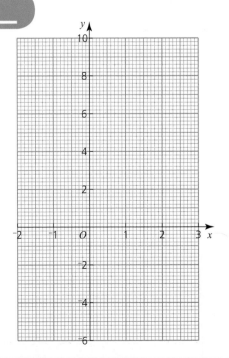

Plot each of the 7 points in the table with a cross (×) to get 1 mark.

Join the 7 points with a smooth curve to get 1 mark.

WARNING

A horizontal line joining (1, −1) and (2, −1) is a common error and would not get full marks since the curve would not then be smooth.

If your plotted points for graphs of quadratic functions of the form $y = ax^2 + bx + c$ do not give a shape like $\cup$ or $\cap$ check your table of values.

The horizontal red line, $y = 3$, is drawn and values of x, where this line crosses the curve, are read off.

b When $y = 3$, $x = -0.55$, $x = 3.55$

Now try these

1 This table of values is for $y = 3x + 1$

x	−2	−1	0	1	2	3
y	−5	−2	1	4	7	10

a On the grid, draw the graph of $y = 3x + 1$

b Use your graph to find
  i the value of y when $x = -0.8$
  ii the value of x when $y = 8.2$

2 This table of values is for $y = 6 - x$

x	−3	−2	−1	0	1	2	3
y	9	8	7	6	5	4	3

a On the grid, draw the graph of $y = 6 - x$
b Use your graph to find the value of x when $y = 7.4$

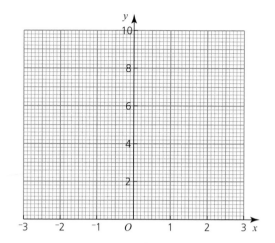

3 This table of values is for $y = 5 - 2x$

x	−2	−1	0	1	2	3	4
y	9	7	5	3	1	−1	−3

a On the grid, draw the graph of $y = 5 - 2x$
b i On the same grid, draw the line $y = 4$
ii Write down the coordinates of the point where your two graphs cross.

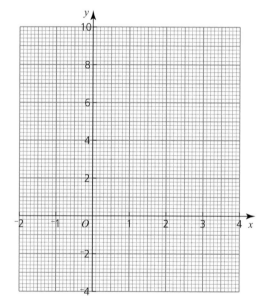

4 This table of values is for $y = x^2 - 4$

x	−3	−2	−1	0	1	2	3
y	5	0	−3	−4	−3	0	5

a On the grid, draw the graph of $y = x^2 - 4$
b What is the minimum value of y?

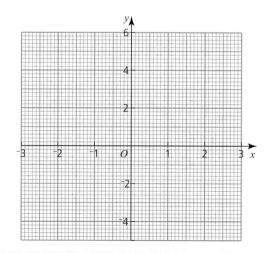

5 This table of values is for $y = x^2 + 2x$

x	−4	−3	−2	−1	0	1	2
y	8	3	0	−1	0	3	8

a On the grid, draw the graph of $y = x^2 + 2x$

b Use your graph to estimate the values of x for which $y = 4$

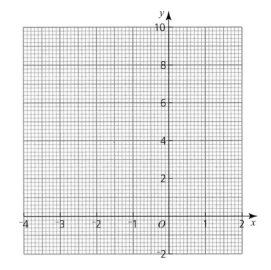

6 This table of values is for $y = 4 + x - x^2$

x	−2	−0	0	1	2	3	4
y	−2	2	4	4	5	−2	−8

a On the grid, draw the graph of $y = 4 + x - x^2$

b Use your graph to find an estimate for the maximum value of y.

c Use your graph to estimate the values of x for which $y = -0.6$

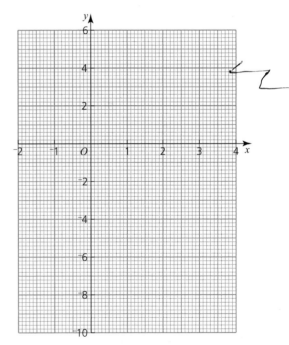

SKILL

Solve linear equations with brackets

KEY FACTS

- To solve linear equations that have brackets, multiply out the brackets; then collect the letter terms on one side of the equals sign and the number terms on the other side.

EXAM FACTS

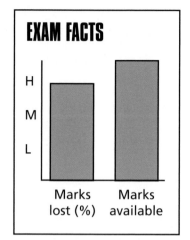

Marks lost (%) | Marks available

REFERENCE

For a reminder of how to collect like terms turn to pages 39 to 40
For a reminder of how to multiply out brackets turn to pages 43 to 44

Getting it right

Solve $3(x + 1) = 8x$

$$3(x + 1) = 8x$$
$$3x + 3 = 8x$$
$$3x + 3 - 3x = 8x - 3x$$
$$3 = 5x$$
$$5x = 3$$

$$x = \frac{3}{5} = 0.6$$

Multiplying out the brackets gets 1 mark.

Subtracting $3x$ from both sides gets 1 mark.
Then collect like terms.
Subtracting $3x$ instead of $8x$ from both sides keeps the x term positive.

$3 = 5x$ is the same as $5x = 3$

Divide both sides by 5

WARNING

A common error is to give the answer

$$x = \frac{5}{3}$$

Solve $5(x + 3) = 2(x + 18)$

$$5(x + 3) = 2(x + 18)$$
$$5x + 15 = 2x + 36$$
$$5x + 15 - 2x = 2x + 36 - 2x$$
$$3x + 15 = 36$$
$$3x = 36 - 15$$
$$3x = 21$$

$$x = \frac{21}{3} = 7$$

Multiplying out one set of brackets correctly gets 1 mark.

Subtracting $2x$ from both sides and then subtracting 15 from both sides gets 1 mark.

Divide both sides by 3

Solve $3(y - 4) = 2 - (y - 1)$

$$3(y - 4) = 2 - (y - 1)$$
$$3y - 12 = 2 - y + 1$$
$$3y - 12 = 3 - y$$
$$3y - 12 + y = 3 - y + y$$
$$4y - 12 = 3$$
$$4y = 3 + 12$$
$$4y = 15$$

$$y = \frac{15}{4} = 3\frac{3}{4} = 3.75$$

Multiply out the brackets and rearrange the terms to get the y terms on one side of the equation and numbers on the other side.

$-(y - 1)$ means $-1 \times (y - 1)$

WARNING

A common error in simplifying $2 - (y - 1)$ is to write
$$2 - y - 1$$
OR to write
$$-2y + 2$$
by multiplying the bracket by -2

The answer can be written as either a fraction or a decimal.

Now try these

In Questions **1-4** solve the equations.

1 $4(x - 1) = 12$ **2** $11 = 5(y + 2)$
3 $15x = 7(2x - 3)$ **4** $3(2 - p) = 5p$

5 Solve $3(x - 4) = x + 24$ *(1387 November 2006)*
6 Solve the equation $6x + 13 = 4(x - 3)$ *(1388 November 2006)*
7 Solve the equation $5(x - 3) = 2x - 22$ *(1388 March 2006)*
8 Solve $1 - 2(1 - x) = 7$
9 Solve $5 - 3x = 2(x + 1)$ *(1388 June 2005)*
10 Solve $4 - (2x - 1) = 2 + x$
11 Solve $6 - 4(2 - 3x) = 2(3 + x)$
12 Solve the equation $3(1 + y) = 3 - (5y + 2)$
13 Solve $6(y - 3) = 3(3y + 2)$
14 Solve the equation $2a - 3(a + 2) = 6(a - 2)$
15 Solve $p + 2(p - 7) = 7 - 6(p + 1)$

SKILL

Derive and solve equations from a diagram

EXAM FACTS

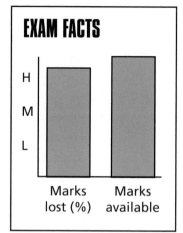

KEY FACTS

- To derive an equation from a diagram use the information given in the diagram.
- This often needs knowledge of geometric facts, for example, the angle sum of a triangle is 180°.
- Then solve the linear equation.

REFERENCE

For a reminder of how to solve equations that have brackets turn to pages 54 to 55

Getting it right

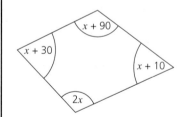

Diagram **NOT** accurately drawn

The sizes, in degrees, of the angles of the quadrilateral are

$x + 10$
$2x$
$x + 90$
$x + 30$

a   Use this information to write down an equation in terms of x.

b   Use your answer to part **a** to work out the size of the smallest angle of the quadrilateral.

EXAM TIP

"Diagram **NOT** accurately drawn" means that taking measurements from the diagram will not give correct answers.

WARNING

A common error is to write these expressions as $10x$, $2x$, $90x$ and $30x$ before attempting to add the expressions.

a $x + 10 + 2x + x + 90 + x + 30 = 360$

$\qquad\qquad\qquad 5x + 130 = 360$

b $\quad 5x + 130 = 360$

$\qquad\qquad 5x = 230$

$\qquad\qquad\;\; x = 230 \div 5$

$\qquad\qquad\;\; x = 46$

So the sizes of the four angles are

$\qquad x + 10 = 56°$

$\qquad\quad\; 2x = 92°$

$\qquad x + 90 = 136°$

$\qquad x + 30 = 76°$

$\qquad$ Smallest angle $= 56°$

The sum of the four angles of a quadrilateral = 360°.

The sum of the angles of this quadrilateral is
$(x + 10) + (2x) + (x + 90) + (x + 30) = 5x + 130$
This gets 1 mark.
Writing this expression equal to 360 gets 1 mark.

Subtract 130 from both sides.

Divide both sides by 5

Substitute $x = 46$ into each of the four expressions.

EXAM TIP

Check that
$56° + 92° + 136° + 76°$ is 360°

Diagram **NOT** accurately drawn

$2x + 3$

$4x + 1$

$6x - 1$

WARNING ⚠️

A common error is to assume that the triangle is equilateral and to write
$2x + 3 = 11 \;(= 33 \div 3)$
$4x + 1 = 11$
$6x - 1 = 11$

All lengths on the diagram are in centimetres.
The perimeter of the triangle is 33 cm.
Work out the value of x.

(1388 November 2006)

Perimeter $= 2x + 3 + 4x + 1 + 6x - 1$

$2x + 3 + 4x + 1 + 6x - 1 = 33$

$\qquad\qquad\qquad 12x + 3 = 33$

$\qquad\qquad\qquad\quad\; 12x = 30$

$x = \dfrac{30}{12} = 2.5$

The perimeter of the triangle is the sum of its three sides. Write each of the expressions for the sides with '+' signs between them.

Put the expression for the perimeter equal to 33

Simplify by collecting like terms and then solve the equation by subtracting 3 from both sides and dividing by 12

EXAM TIP

Always check your answer by substituting. Putting $x = 2.5$ into each expression on the diagram gives 8, 11 and 14 $(8 + 11 + 14 = 33)$

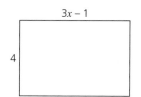

Now try these

1   The diagram shows a rectangle.

All lengths on the diagram are in centimetres.

The perimeter of the rectangle is 24 cm.

Diagram **NOT** accurately drawn

   a   Use this information to write down an equation in terms of x.
   b   Find the length of the longer side of the rectangle.

2   The diagram shows an isosceles triangle.
    All angles on the diagram are in degrees.

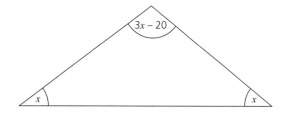

Diagram **NOT**
accurately drawn

   a   Use this information to write down an equation in terms of x.
   b   By solving your equation, find the size of the obtuse angle.

3   In the diagram, all measurements are in centimetres.
    ABC is an isosceles triangle.
    $AB = 2x$
    $AC = 2x$
    $BC = 10$

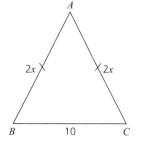

Diagram **NOT**
accurately drawn

   a   Find an expression, in terms of x, for the **perimeter**
       of the triangle.
       Simplify your expression.

   The perimeter of the triangle is 34 cm.
   b   Find the value of x.

(1387 June 2006)

4 The diagram shows two blocks.
The mass, in kilograms, of each block is
shown on the diagram.

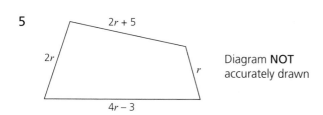

Diagram **NOT**
accurately drawn

The masses of the two blocks are equal.

a Use this information to write down
an equation in terms of m.
b Solve your equation to find the value of m.
c Find the mass of each block.

5

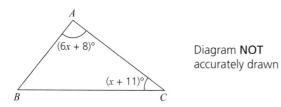

Diagram **NOT**
accurately drawn

In the diagram, all measurements are in centimetres.
The lengths of the sides of the quadrilateral are

$2r + 5$
$2r$
$4r - 3$
r

a Find an expression, in terms of r, for the perimeter of the quadrilateral.
Give your expression in its simplest form.

The perimeter of the quadrilateral is 65 cm.
b Work out the value of r.   *(1387 November 2006)*

6

A
$(6x + 8)°$
$(x + 11)°$
B   C

Diagram **NOT**
accurately drawn

ABC is a triangle.
Angle $A = (6x + 8)°$
Angle $C = (x + 11)°$

The size of angle A is **four** times the size of angle C.

a Work out the value of x.
b Work out the size of angle B.

SKILL

Change the subject of a formula

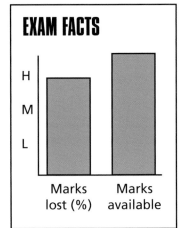
KEY FACTS

- The subject of a formula appears just once, and only on the left-hand side of the formula.
- To change the subject of a formula rearrange the terms of the formula.
- For example,
  v is the subject of the formula $v = u + 5t$.
- To make t the subject of this formula, rearrange the terms to give $t = \ldots$

Getting it right

Make t the subject of the formula
$$v = u + 5t$$

(1388 June 2005)

Method 1
$$v = u + 5t$$
$$v - u = 5t$$
$$\frac{v - u}{5} = t$$
$$t = \frac{v - u}{5}$$

> Subtracting u from both sides gets the term with t on its own. This gets 1 mark.

> Divide both sides by 5 to get $t = \ldots$

WARNING

A common error is to write
$$t = v - u/5$$
This would not get full marks because
$$t = v - u/5 \quad \text{means} \quad t = v - \frac{u}{5}$$

WARNING

Method 2 should not be used if t occurs in more than one term in the given formula.

Method 2

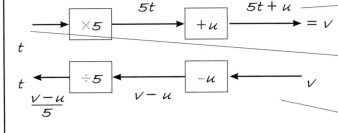

$$t = \frac{v - u}{5}$$

> To get $v = u + 5t$ multiply t by 5 and then add u

> Start with t since t is to be the subject.

> Reverse the order, starting with v, and use inverse operations. Subtract u and then divide by 5

Rearrange

$$a(q - c) = d$$

to make q the subject.

(1387 November 2005)

$$a(q - c) = d$$

$$aq - ac = d$$

$$aq = d + ac$$

$$q = \frac{d + ac}{a}$$

Multiplying out the brackets gets 1 mark.

Adding ac to both sides to get the term with q on its own gets 1 mark.

Divide both sides by a to get $q = ...$

WARNING

$\frac{d + ac}{a}$ cannot be simplified.

A common error is to cancel the as and this would lose a mark.

Make a the subject of the formula

$$r = \frac{2(a + b)}{c}$$

$$r = \frac{2(a + b)}{c}$$

$$r = \frac{2a + 2b}{c}$$

$$rc = 2a + 2b$$

$$rc - 2b = 2a$$

$$\frac{rc - 2b}{2} = \frac{2a}{2}$$

$$\frac{rc - 2b}{2} = a$$

$$a = \frac{rc - 2b}{2}$$

Multiply out the bracket.

Multiplying both sides by c to remove the fraction gets 1 mark.

WARNING

A common error when multiplying both sides by c is to try to multiply the numerator as well as the denominator by c.

Get the term with a on its own by subtracting $2b$ from both sides.

Divide both sides by 2 to get $a = ...$

WARNING

A common error is to cancel the 2s to get the wrong answer

$$a = rc - b$$

1 Make c the subject of the formula $f = 3c - 4$ (*1388 November 2005*)

2 Make p the subject of the formula $c = 5p - 2$ (*1388 November 2006*)

3 Make r the subject of the formula $p = 3 - 4r$

4 Make b the subject of the formula $P = 2a + 2b$ (*1388 March 2006*)

5 Rearrange $a(b + c) = d$ to make b the subject.

6 Rearrange $x(6 - y) = 4z$ to make y the subject.

7 Make m the subject of the formula $y = 2(m + 1) + 3(m - 2)$

8 Rearrange $P = \dfrac{r + 3}{5}$ to make r the subject.

9 Make B the subject of the formula $D = \dfrac{A(B + 2)}{C}$

10 Rearrange $\dfrac{x}{2} = \dfrac{1}{y}$ to make

   a x the subject,

   b y the subject.

SKILL

Find the nth term of a linear number sequence

KEY FACTS

- To find the nth term of an arithmetic sequence use
  nth term is difference $\times$ n + zero term

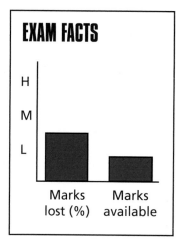
Getting it right

The first five terms of an arithmetic sequence are

| 3 | 8 | 13 | 18 | 23 |

Find, in terms of n, an expression for the nth term of this sequence.

$$3 \quad 8 \quad 13 \quad 18 \quad 23$$
$$+5 \quad +5 \quad +5 \quad +5$$

Difference $= +5$

First term $= 3$

Zero term $= 3 - 5 = -2$

nth term is $5 \times n + (-2)$

nth term is $5n - 2$

WARNING

A common error is to just write the answer as $n + 5$

Difference between consecutive terms is $8 - 3 = 5$, $13 - 8 = 5$, and so on.

The zero term is the term before the first term (3).

nth term is difference $\times$ n + zero term

Complete the table.

Position of term	Number sequence
1st	11
2nd	20
3rd	29
4th	38
5th	
nth	

WARNING

Writing the answer as $n = 5n - 2$ would lose a mark.

Position of term	Number sequence
1st	11
2nd	20
3rd	29
4th	38
5th	47
nth	$9n + 2$

$+9$
$+9$
$+9$
$+9$

Difference between consecutive terms is $+9$, so add 9 to the 4th term to find the 5th term.

WARNING

A common error is to just add 9 to the 5th term to get nth term.

nth term is **not** 56

Difference $= +9$

So 5th term $= 38 + 9 = 47$

Zero term $= 11 - 9 = +2$

nth term is $9n + 2$

nth term is difference $\times n$ + zero term

The first five terms of an arithmetic sequence are

5   3   1   −1   −3

a Find, in terms of n, an expression for the nth term of this sequence.

b Adeel says that -122 is a term in this sequence.
Is Adeel correct?
You must justify your answer.

WARNING

A common error is to just write the difference as 2 but since this sequence is decreasing the difference must be negative 2

a   5   3   1   −1   −3
        −2   −2   −2   −2

Difference between consecutive terms is $3 - 5 = -2$, $1 - 3 = -2$, and so on.

Difference $= -2$

Zero term $= 5 + 2 = 7$

The zero term is the term before the first term.

nth term is $-2n + 7$

nth term is difference $\times n$ + zero term

b -122 is **not** a term in this sequence because the number part of each term in the sequence is always odd.

Now try these

1   The first five terms of an arithmetic sequence are

     8   11   14   17   20

Find, in terms of n, an expression for the nth term of this sequence.

2   Here are some patterns made from sticks.

Complete the table.

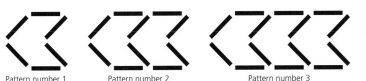

Pattern number 1   Pattern number 2   Pattern number 3

Pattern number	Number of sticks
1	6
2	10
3	14
4	18
5	
n	

(1388 June 2006)

3   Here are the first five terms of a number sequence.

     3   7   11   15   19

a   Work out the **8th** term of the number sequence.

b   Write down an expression, in terms of n, for the nth term of the number sequence.

(1387 November 2006)

4   Here are the first four terms of a number sequence.

     2   7   12   17

a   Write down the **10th** term of this number sequence.

b   Write down an expression, in terms of n, for the nth term of this number sequence.

c   The nth term of a different number sequence is $2n$.
Find the first three terms of this number sequence.

5   Here are the first six terms of an arithmetic sequence.

17   12   7   2   −3   −8

Write down, in terms of n, an expression for the nth term of this sequence.

6   Here are the first five terms of an arithmetic sequence.

−14   −7   0   7   14

a   Write down, in terms of n, an expression for the nth term of this sequence.

b   Rebecca says that 756 is a term in this sequence. Is Rebecca correct?
You must justify your answer.

23 Trial and improvement

SKILL	EXAM FACTS

SKILL

Solve equations by trial and improvement

KEY FACTS

- Approximate solutions to equations which cannot be solved exactly can be found by first guessing the solution and then using the method of trial and improvement to obtain a more accurate answer.

EXAM FACTS

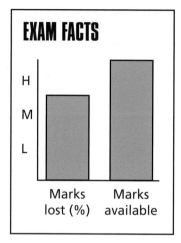

Getting it right

a   Show that the equation $x^3 - 7x = 68$ has a solution between 4 and 5.

b   Use a trial and improvement method to find this solution. Give your answer correct to one decimal place. You must show **all** your working.

WARNING

In this type of question giving the correct answer without showing any working will get no marks.

$x = 4$ is too small and $x = 5$ is too big so the solution is between $x = 4$ and $x = 5$

a

x	LHS $x^3 - 7x$	Compare with RHS 68	
4	$4^3 - 7 \times 4 = 36$	Since 36<68, $x=4$ is too small	
5	$5^3 - 7 \times 5 = 90$	Since 90>68, $x=5$ is too big	Solution is between 4 and 5

b

x	LHS $x^3 - 7x$	Compare with RHS 68	
4.5	$4.5^3 - 7 \times 4.5$ $= 59.625$	$59.625 < 68$, so $x = 4.5$ is too small	
4.6	$4.6^3 - 7 \times 4.6$ $= 65.136$	$65.136 < 68$, so $x = 4.6$ is too small	
4.7	$4.7^3 - 7 \times 4.7$ $= 70.923$	$70.923 > 68$, so $x = 4.7$ is too big	Solution is between 4.6 and 4.7
4.65	$4.65^3 - 7 \times 4.65$ $= 67.994625$	$67.994625 < 68$ so $x = 4.65$ is too small	Solution is between 4.65 and 4.7

Correct to one decimal place, the solution is $x = 4.7$

Try $x = 4.5$ since the solution lies between 4 and 5

The value 59.625 must be shown to get 1 mark.

Since $x = 4.6$ is still too small, try $x = 4.7$

The value 70.923 shows that $x = 4.7$ is too big. This gets 1 mark.

To decide if the solution is closer to 4.6 or 4.7, you MUST find the middle value, 4.65, of the interval.

WARNING

A common error is to write the solution as 67.994... but you are looking for the value of x not $x^3 - 7x$.

WARNING

A common error is to write the solution as $x = 4.65$ instead of giving the solution to 1 decimal place.

Use a trial and improvement method to find the solution of $x^2 = \dfrac{2}{x} + 5$ that lies between 2 and 3. Give your answer correct to one decimal place. You must show **all** your working.

x	LHS x^2	RHS $\dfrac{2}{x} + 5$	Compare LHS and RHS
2.5	$2.5^2 = 6.25$	$\dfrac{2}{2.5} + 5 = 5.8$	Since $6.25 > 5.8$ $x = 2.5$ is too big
2.4	$2.4^2 = 5.76$	$\dfrac{2}{2.4} + 5$ $= 5.83...$	Since $5.76 < 5.83...$ $x = 2.4$ is too small
2.45	2.45^2 $= 6.0025$	$\dfrac{2}{2.45} + 5$ $= 5.816...$	Since $6.0025 > 5.816...$ $x = 2.45$ is too big

So, correct to one decimal place, the solution is $x = 2.4$

Work out the value of the left-hand side (LHS).

Work out the value of the right-hand side (RHS) and compare it with the value of the left-hand side (LHS).

```
2.4          2.45          2.5
              ↑
           too big
```

1 Show that each of these equations has a solution between 2 and 3

   a $x^3 - 4x = 2$ b $x^3 = 8x - 1$

2 The equation

   $$x^3 - x = 140$$

   has a solution between 5 and 6
   Use a trial and improvement method to find this solution.
   Give your solution correct to 1 decimal place.
   You must show **all** your working.

   (1388 November 2006)

3 The equation

   $$x^3 + 4x^2 = 100$$

   has a solution between 3 and 4
   Use a trial and improvement method to find this solution.
   Give your solution correct to one decimal place.
   You must show **all** your working.

   (1387 November 2006)

4 a Show that the equation $x^3 - 5x = 27$ has a solution between 3 and 4
   b Use a trial and improvement method to find this solution.
      Give your answer correct to one decimal place.
      You must show **all** your working.

5 The equation

   $$x(x^2 - 6) = 60$$

   has a solution between 4 and 5
   Use a trial and improvement method to find this solution.
   Give your solution correct to 1 decimal place.
   You must show **all** your working.

6 The equation

   $$x^2 = \frac{4}{x} + 13$$

   has a solution between 3 and 4
   Use a trial and improvement method to find this solution.
   Give your solution correct to 1 decimal place.
   You must show **all** your working.

7 Use a trial and improvement method to find the solution of

   $$x^3 = x + 18$$

   Give your solution correct to 1 decimal place.
   You must show **all** your working.

24 Area of a triangle

KEY FACTS

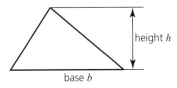

base b

height h

- To work out the area of a triangle, work out a half of its base times its height *or* multiply the length of the base by the height and divide the result by 2

- That is

$$\text{Area} = \frac{1}{2} \times \text{base} \times \text{height}$$

or $\text{Area} = \frac{1}{2}\text{base} \times \text{height}$

or $\text{Area} = \dfrac{\text{base} \times \text{height}}{2}$

$A = \frac{1}{2}bh$

- Remember that the height of a triangle is the vertical height or the perpendicular height.

EXAM FACTS

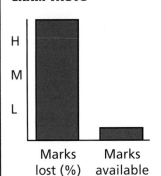

| | Marks lost (%) | Marks available |

EXAM TIP

"Diagram **NOT** accurately drawn" means that taking measurements from the diagram will not give correct answers.

EXAM TIP

Questions on working out the area of a triangle could be on either paper. This question could have been on a non-calculator paper.

WARNING

A common error is to forget to divide by 2
half of 6 is 3, that is $\frac{1}{2} \times 6 = 3$

WARNING

Remember to write the units if they are not given on the answer line.

Getting it right

Calculate the area of the triangle.

Diagram **NOT** accurately drawn

10 cm

6 cm

Use
$$A = \frac{1}{2} \times \text{base} \times \text{height}$$

Show your working. You would get 1 mark for this.

$\text{Area} = \frac{1}{2} \times 6 \times 10$

$\text{Area} = 3 \times 10$

$= 30 \text{ cm}^2$

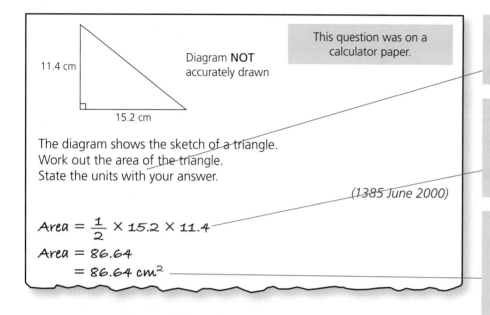

11.4 cm

15.2 cm

Diagram **NOT** accurately drawn

This question was on a calculator paper.

The diagram shows the sketch of a triangle.
Work out the area of the triangle.
State the units with your answer.

(1385 June 2000)

Units would not be printed on the answer line. You have to give the units of your answer.

To work out $\frac{1}{2} \times 15.4 \times 11.2$ on your calculator, you can find 15.4×11.2 and then divide by two or you can just key in $0.5 \times 15.4 \times 11.2$

Area $= \frac{1}{2} \times 15.2 \times 11.4$

Area $= 86.64$

$= 86.64$ cm^2

The question does not ask you to give your answer to any particular degree of accuracy so write down all the figures on your calculator display. Do not forget to give the units of the area. You would get 1 mark for this even if your answer to the calculation is wrong.

Now try these

The diagrams are not accurately drawn.
In Questions **1–8**, work out the area of each triangle and give the units of your answer.

1

3 cm

6 cm

2

8 cm

5 cm

3

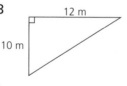

12 m

10 m

4

6 cm

14 cm

5

6.3 cm

8.7 cm

6

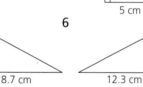

8.5 cm

12.3 cm

7

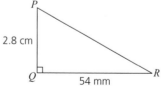

68 mm

21 mm

8

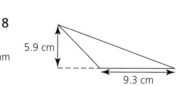

5.9 cm

9.3 cm

9 *PQR* is a right-angled triangle.
Work out the area of the triangle in
  a mm^2,   **b** cm^2.

P

2.8 cm

Q   54 mm   R

10 Triangle *ABC* is a right-angled triangle.
AB = 12 cm, *AC* = 5 cm, *BC* = 13 cm
Angle *A* = 90°
  a Work out the area of triangle *ABC*.
      D is the point on *BC* such that *AD* is perpendicular to *BC*.
  b Work out the length of *AD*.
      Give your answer correct to 2 decimal places.

5 cm

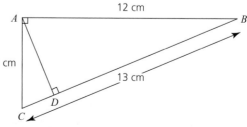

12 cm

13 cm

A   B

C   D

SKILL

Find the area and perimeter of compound shapes

EXAM FACTS

H
M
L

Marks lost (%) Marks available

KEY FACTS

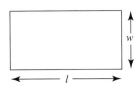

- To work out the perimeter of a **rectangle** add the lengths of the four sides or add the length and the width and then multiply the result by 2
  $P = l + w + l + w = 2l + 2w = 2(l + w)$

- To work out the area of a rectangle multiply the length by the width
  $A = l \times w$ or $A = lw$

- To work out the area of a **parallelogram** multiply the length of the base by the vertical height $A = b \times h = bh$

- To work out the area of a **trapezium** add the lengths of the two parallel sides, multiply the result by the height and then divide the result by 2

  $A = \frac{1}{2}(a + b)h$

EXAM TIP

This formula is on the Foundation tier exam formula sheet, but not on the Higher.

- Compound shapes are made from simpler shapes.

- To find the perimeter of a compound shape add the lengths of the sides of the shape.

- To find the area of a compound shape split the shape into parts that you can work out the area of and add the areas of the parts.

REFERENCE

For a reminder of how to find the area of a triangle, turn to pages 69 to 70

Getting it right

The diagram shows a shape.
The shape is a 6-sided polygon.
The diagram shows the lengths of two of the sides of the shape.
Work out the perimeter of the shape. *(1387 June 2005)*

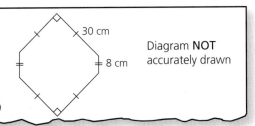

30 cm

8 cm

Diagram **NOT** accurately drawn

EXAM TIP

"Diagram **NOT** accurately drawn" means that taking measurements from the diagram will not give correct answers.

Perimeter = 4 × 30 + 2 × 8
         = 120 + 16 = 136
Perimeter = 136 cm

The shape has 4 sides of length 30 cm and 2 sides of length 8 cm.
The perimeter can be worked out by adding the lengths of all 6 sides
30 + 30 + 30 + 30 + 8 + 8
but it is easier to write this as 4 × 30 + 2 × 8

This diagram shows the floor plan of a room.

6 m

Diagram **NOT** accurately drawn

6 m

4 m

11 m

Work out the area of the floor. Give the units with your answer.

Split the diagram into two parts: a square and a rectangle.
You would get one mark for this. (There are other ways to split up the diagram.)

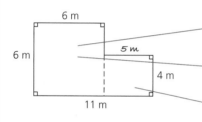

6 m

6 m

5 m

4 m

11 m

This is a square of side 6 m.

Area of square = 6 × 6 = 36m²
Area of rectangle = 5 × 4 = 20m²
Area of the floor = 36 + 20 = 56 m²

This is a rectangle. The length has to be worked out. The length = (11 − 6) = 5m. The width is 4 m.

Give the units with your answer, as the question asks. You can get a mark for this even if the area is wrong.

Bill has a rectangular sheet of metal. The length of the rectangle is 12.5 cm. The width of the rectangle is 10 cm.

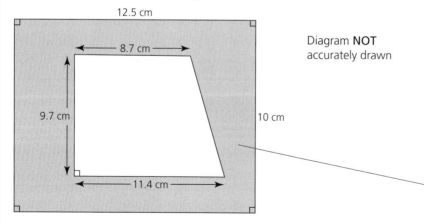

12.5 cm

8.7 cm

Diagram **NOT** accurately drawn

9.7 cm

10 cm

11.4 cm

To work out the area of the shaded region take the area of the trapezium away from the area of the rectangle.

Bill cuts out a trapezium. The dimensions of the trapezium are shown in the diagram. He throws away the rest of the metal sheet.

Work out the area of the metal sheet, shown shaded in the diagram, that he throws away. Give your answer in cm², correct to 1 decimal place.

Area of rectangle = 12.5 × 10 = 125cm²
Area of trapezium = $\frac{1}{2}$ (8.7 + 11.4) × 9.7
$\qquad\qquad\qquad = \frac{1}{2}$ × 20.1 × 9.7
$\qquad\qquad\qquad = 97.485$ cm²

Area of shaded region = 125 − 97.485 = 27.515 cm²

Area of metal sheet thrown away = 27.5 cm²

To work out the area of the rectangle use $A = l \times w$ with $l = 12.5$ and $w = 10$

To work out the area of the trapezium use
$A = \frac{1}{2}(a + b)h$ with
$a = 8.7$, $b = 11.4$, $h = 9.7$

WARNING

Use a calculator and work out 8.7 + 11.4 first before you multiply as these are in brackets.

Now try these

EXAM TIP

Work to more than 1 decimal place so that you can correct your final answer to 1 decimal place.

1 Work out the area of the shape.

Diagram **NOT** accurately drawn

5 m

7 m

13 m

(1385 November 2001)

2 The diagram shows a trapezium of height 3 m.
   Find the area of this trapezium.
   State the units with your answer.

Diagram **NOT** accurately drawn

2 m

3 m

6 m

(1387 November 2005)

3 The diagram shows a paved surface.
   All the corners are right angles.
   Work out the area of the paved surface.

Diagram **NOT** accurately drawn

12 m

7 m

3 m

6 m

2 m

2 m

(1385 June 2002)

4 Work out the perimeter of the paved surface in Question 3.

5 The diagram shows 3 small rectangles inside a large rectangle.
The large rectangle is 10 cm by 8 cm.
Each of the 3 small rectangles is 4 cm by 2 cm.
Work out the area of the region shown shaded in the diagram.

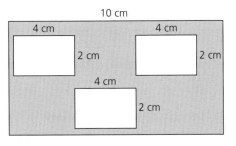

Diagram **NOT** accurately drawn

(1387 June 2006)

6 Here is a side view of a swimming pool.
ABCD is a horizontal straight line. *AH*, *BG*, *CF* and *DE* are vertical lines.

a Write down the mathematical name for the quadrilateral *BCFG*.
b Work out the area of quadrilateral *BCFG*.
c Hence work out the area of the hexagon *ADEFGH*.
Give your answer in m² correct to 1 decimal place.

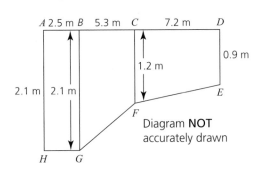

Diagram **NOT** accurately drawn

(1384 June 1996)

7 The diagram shows a rectangle inside a triangle.
The triangle has a base of 12 cm and a height of 10 cm.
The rectangle is 5 cm by 3 cm.
Work out the area of the region shown shaded in the diagram.

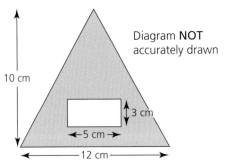

Diagram **NOT** accurately drawn

8 The diagram shows a shape.
All the corners of the shape are right angles.
Work out the perimeter of the shape.

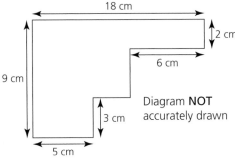

Diagram **NOT** accurately drawn

9 The diagram shows a car park. Mrs Roberts is selling the car park. She will accept any offer that is more than £28 per square metre.
Mr Patel offers £194 700 for the car park.
Will Mrs Roberts accept Mr Patel's offer for the car park?
You must show how you reached your decision.

(1387 Mock paper)

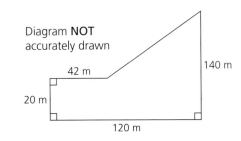

Diagram **NOT** accurately drawn

26 Angle sums of triangles and quadrilaterals

SKILLS

Use the sum of the angles of a triangle = 180°
to find a missing angle

Use the sum of the angles of a quadrilateral = 360°
to find a missing angle

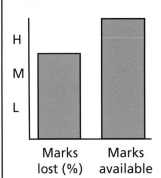
KEY FACTS

* The angle sum of a triangle is 180°.
  $a + b + c = 180°$

* The angle sum of a quadrilateral is 360°.
  $a + b + c + d = 360°$

* Any quadrilateral can be split into two triangles and so the angle sum of a quadrilateral is double the angle sum of a triangle.

Getting it right

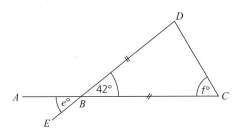

Diagram **NOT** accurately drawn

ABC and EBD are straight lines.
$BD = BC$. Angle $CBD = 42°$.

a   Write down the size of the angle marked $e°$.
b   Work out the size of the angle marked $f°$.

(1385 November 2000)

a $e° = 42°$

b $42 + f + f = 180$
    $2f = 180 - 42$
    $f = \dfrac{138}{2}$
    $f° = 69°$

The reason is 'Where two straight lines cross, the opposite angles are equal'. Sometimes, reasons are asked for.

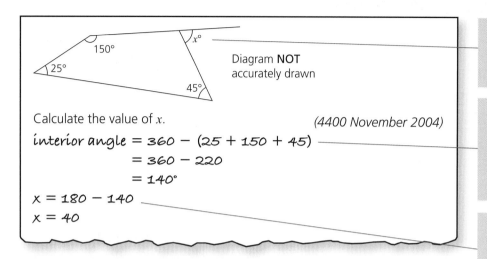

150°

25°

$x°$

45°

Diagram **NOT** accurately drawn

Calculate the value of x.

(4400 November 2004)

interior angle $= 360 - (25 + 150 + 45)$

$= 360 - 220$

$= 140°$

$x = 180 - 140$

$x = 40$

Three of the interior angles of the quadrilateral are given but an **exterior** angle is required.

Show how to use the angle sum of a quadrilateral to find the remaining interior angle. You would get a mark for this, even if you worked it out wrongly.

$x°$ and 140° are angles on a straight line and the sum of the angles on a straight line is 180°.

Now try these

The diagrams are not accurately drawn.
In Questions **1–8**, find the size of each of the angles marked with letters and show your working.

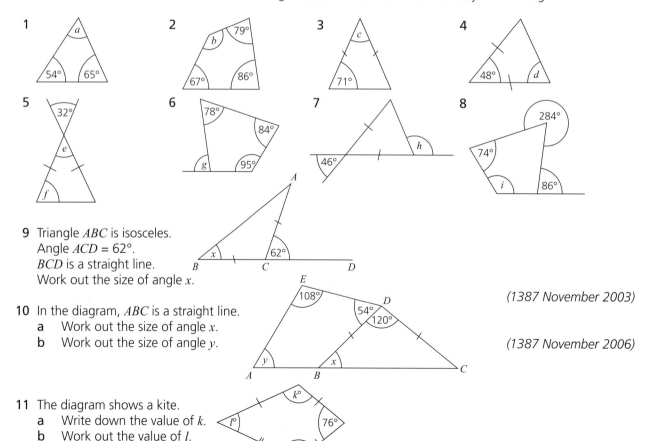

1 a 54°   65°

2   79° b 67°   86°

3 c 71°

4   48° d

5   32° e f

6   78°   84° g 95°

7   46° h

8   284°   74° i 86°

9 Triangle ABC is isosceles.
 Angle $ACD = 62°$.
 BCD is a straight line.
 Work out the size of angle x.

(1387 November 2003)

10 In the diagram, ABC is a straight line.
 a   Work out the size of angle x.
 b   Work out the size of angle y.

(1387 November 2006)

11 The diagram shows a kite.
 a   Write down the value of k.
 b   Work out the value of l.

A x 62° B C D

E 108° D 54°   120° y x A B C

$k°$ $l°$ 76°   118°

27 Regular polygons – interior and exterior angles

SKILLS

Find the sizes of interior and exterior angles of a regular polygon

Find the number of sides of a regular polygon, given the size of an exterior angle

EXAM FACTS

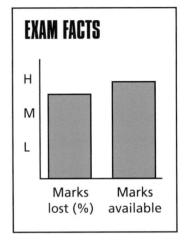

KEY FACTS

- A **polygon** is a shape with three or more straight sides. You should to know these special names: pentagon – 5 sides, hexagon – 6 sides, octagon – 8 sides, decagon – 10 sides

- A polygon's **interior** angles are the angles *inside* the polygon.

- To find the sum of the interior angles of a polygon, subtract 2 from the number of sides and multiply the result by 180°.
  So, for an *n*-sided polygon, the sum of the interior angles is $(n - 2) \times 180°$.

- At a vertex,
  interior angle + exterior angle = 180°
  (because two angles on a straight line add to 180°).

- The sum of the exterior angles of **any** polygon is 360°.

- A **regular** polygon has all its sides the same length and all its angles the same size.

- The size of each exterior angle of a regular *n*-sided polygon is $\frac{360°}{n}$

Getting it right

Work out the size of each interior angle of a regular octagon. *(1388 January 2003)*

Diagram **NOT** accurately drawn

Method 1
Size of each exterior angle = 360° ÷ 8 = 45°
Size of each interior angle = 180° − 45° = 135°
Method 2
Sum of interior angles = (8 − 2) × 180°
                      = 6 × 180°
                      = 1080°

EXAM TIP

"Diagram **NOT** accurately drawn" means that taking measurements from the diagram will not give correct answers.

WARNING

A common mistake is to use 180° as the sum of the exterior angles.

Substitute *n* = 8 into $(n - 2) \times 180°$

$$\text{Size of each interior angle} = 1080° \div 8$$
$$= 135°$$

Method 1 is simpler if the size of **each** interior angle is required but Method 2 is probably better for finding the sum of the interior angles.

The size of each exterior angle of a polygon is 30°. Work out the number of sides of the polygon. *(1388 January 2003)*

Number of sides = 360 ÷ 30
= 12

As all the exterior angles of the polygon are equal, the polygon is regular.

Now try these

The diagrams are not accurately drawn.

1 Here is a regular polygon with 9 sides.
Work out the size of the angle marked $a°$.

2 Work out the size of an exterior angle of a regular pentagon.

(1387 June 2004)

3 Work out the size of an exterior angle of a regular 30-sided polygon.

4 Work out the size of an exterior angle of a regular decagon.

(1388 March 2005)

5 The diagram shows a regular hexagon.
  a Work out the size of angle x.
  b Work out the size of angle y.

(1388 November 2006)

6 The diagram shows a regular
12-sided polygon.
Work out the size of the angle marked $x°$.

7 Work out the size of each interior angle of a regular 36-sided polygon.

8 Work out the number of sides of a regular polygon
  a if each exterior angle is 15°   b if each exterior angle is 9°.

9 The size of each interior angle of a regular polygon is 156°.
Work out the number of sides of the polygon.   *(1388 January 2005)*

10 The size of each interior angle of a regular polygon is 160°.
Work out the number of sides of the polygon.

11 The size of each interior angle of a regular polygon is 162°.
Work out the sum of the interior angles of the polygon.

SKILL

Measure a bearing

KEY FACTS

- Bearings are measured **clockwise** ↺ from **North**.
- When the angle is less than 100°, one or two zeros are written in front of the angle, so that the bearing still has three figures. They are often called 'three figure bearings.'

EXAM FACTS

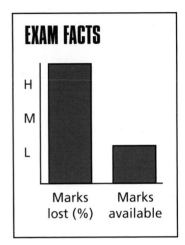

Getting it right

This is a map of part of Northern England.

Measure and write down the bearing of

a Halifax from Wigan,
b Preston from Manchester.

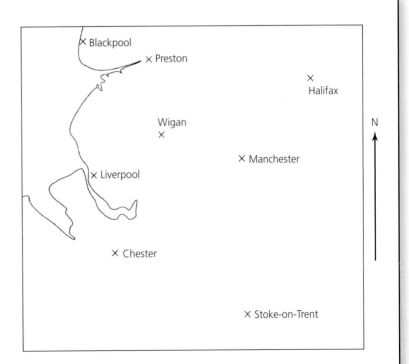

(1387 June 2004)

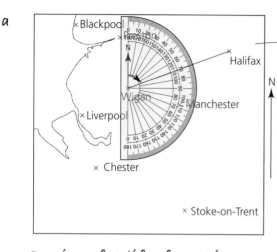

a

Bearing of Halifax from Wigan = 070°

Draw a North line at Wigan. Draw a line from Wigan to Halifax. Measure the acute angle between the lines.

EXAM TIP

Any answer in the range 068°–072° would be accepted.

WARNING

Write a zero in front of the 70° so that the bearing has three figures.

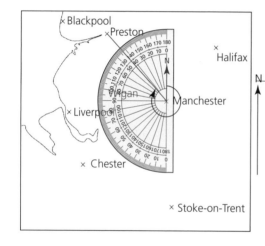

b

Acute angle = 44°
Bearing of Preston from Manchester = 360° − 44°
= 316°

Draw a North line at Manchester.
Draw a line from Manchester to Preston.
The required bearing is the **reflex** angle between the lines.
Unless you are using a circular protractor, measure the acute angle between the lines and subtract it from 360°.

Now try these

1   Use the map on page 79 to measure the bearing of
   a   Halifax from Stoke-on-Trent   b   Chester from Preston
   c   Liverpool from Blackpool   d   Preston from Halifax

2   Use the map on page 79 to find the town which is on a bearing of
   a   179° from Manchester   b   026° from Liverpool
   c   319° from Wigan   d   230° from Halifax

3 Measure and write down the bearing
of the point B from the point A.

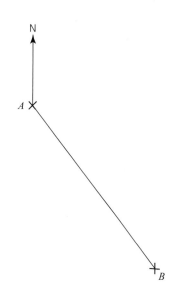

(1385 November 2001)

4 By measurement, find the bearing
of B from A.

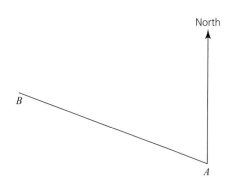

(4400 November 2006)

5 The grid represents part of a map.
Copy the diagram onto centimetre
squared paper.
  a On your grid, draw a line on a
  bearing of 037° from the point
  marked A.
The point C is on a bearing of 300°
from the point marked B.
C is also 3 cm from B.
  b Mark the position of the point C
  on the grid and label it with the
  letter C.

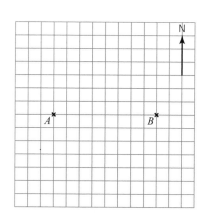

(1385 May 2002)

6 N

A   B

The diagram shows 2 points, A and B.
Copy the diagram.
C is a point such that the bearing of C from A is 041° and the bearing
of C from B is 294°.
On your copy of the diagram, make an accurate drawing to find the
position of the point C.

7 The diagram shows the position of a farm F
and a bridge B on a map.

    a Measure and write down the bearing
       of B from F.

Copy the diagram.
A church C is on a bearing of 155°
from the bridge B.
On the map, the church is 5 cm from B.

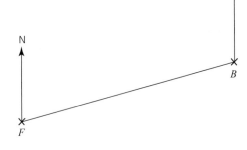

    b On your copy of the diagram, mark the church
       with a cross (X) and
       label it C.

(1387 November 2006)

SKILL

Use *average speed = distance ÷ time*
in calculations

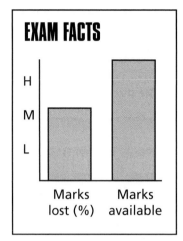

KEY FACTS

- With D for distance, S for average speed and T for time, the diagram can be used as a reminder that

    distance = average speed × time

    $$\text{average speed} = \frac{\text{distance}}{\text{time}}$$

    $$\text{time} = \frac{\text{distance}}{\text{average speed}}$$

- In a calculation the units must always agree.
  For example, km/h, km and hours not km/h, m and minutes.

Getting it right

Sally drove from Birmingham to Manchester in 2 hours.
She drove at an average speed of 44 mph.
Work out the distance that Sally drove.

distance = 44 × 2
        = 88
Sally drove 88 miles.

> mph means miles per hour

> Use distance = average speed × time

> Show your working. You would get 1 mark for this.

> Speed is in miles per hour and time is in hours so the distance is in miles.

Bridget flew from the UK to Dubai.
Bridget's flight from the UK to Dubai covered a distance of 5456 km.
The flight time was 7 hours 45 minutes.
Work out the average speed of the flight.

(4400 November 2006)

45 minutes = $\frac{45}{60}$ = $\frac{3}{4}$ = 0.75 hours

> Use average speed = $\frac{\text{distance}}{\text{time}}$

> Show your working.

so 7 hours 45 minutes = 7.75 hours

average speed = $\frac{5456}{7.75}$ = 704

The average speed of the flight = 704 km/h

WARNING

A common error is to write 7 hours 45 minutes as 7.45 hours.

You would get 1 mark for 7.75

Now try these

1 An aeroplane flies from Liverpool to Prague, a distance of 1200 km. The aeroplane takes 4 hours. Work out the average speed of the aeroplane. State the units of your answer.   *(1387 June 2006)*

HINT

The distance is in km and the time is in hours, so the speed is in km/h.

2 Simon went for a cycle ride. His average speed for the ride was 25 km/h. Simon rode for 4 hours. Work out the distance that Simon rode.

3 Mia drove a distance of 343 km. She took 3 hours 30 minutes. Work out her average speed. Give your answer in km/h.   *(1388 June 2003)*

4 The distance from Bristol to Hull is 330 km. Trevor drove from Bristol to Hull at an average speed of 60 km/h. Work out the time, in hours, his journey took.

5 Suhail cycles 117 km in 4 hours 30 minutes. Work out his average speed.   *(4400 May 2004)*

6 Connor drove from his home to visit a friend. The journey to his friend's house took 2 hours 15 minutes. Connor drove at an average speed of 56 km/h.
   a Work out the distance that Connor drove from his home to his friend's house.
   On his way home Connor had to take a different route because of roadworks. His journey home was 9 km longer than his journey to his friend's house and he drove at an average speed of 54 km/h.
   b Work out how long, in hours and minutes, the journey home took Connor.

7 The distance from Bristol to Leeds is 216 miles. Cara drove the 216 miles in 4 hours 30 minutes. Calculate her average speed. State the units of your answer.   *(1387 November 2006)*

8 Nilesh drove a distance of 84 km at an average speed of 63 km/h. He started his journey at 10 00. At what time did he finish his journey?

9 The distance from Glasgow to Edinburgh is 72 km. James travelled from Glasgow to Edinburgh in 1 hour 36 minutes. Work out James' average speed.

10 Diane drove a distance of 65 km in 50 minutes. She stopped to eat for 25 minutes and then drove a further distance of 45 km in 60 minutes. Work out Diane's average speed for the whole journey. Give your answer in km/h, correct to 1 decimal place.

SKILLS

Use the formulae $C = \pi d$ or $C = 2\pi r$ to find the circumference of a circle

Use the formula $A = \pi r^2$ to find the area of a circle

KEY FACTS

- π is a number. Use your calculator's π button, if it has one. Otherwise use $\pi = 3.142$

- To find the circumference of a circle, multiply the diameter by π.

  $C = \pi d$ or $C = 2\pi r$

- To find the area of a circle, multiply the square of the radius by π.

  $A = \pi r^2$ or $A = \pi \times r \times r$

Getting it right

The (diameter) of a circle is 12 centimetres.
Work out the (circumference) of the circle.
Give your answer correct to 1 decimal place.   *(1387 November 2004)*

$c = \pi \times 12$
$\quad = 37.699...$
Circumference $= 37.7$ cm (to 1 d.p.)

The diameter is given. The circumference is required. So use $C = \pi d$

Show your working. You would get 1 mark for this.

EXAM TIP

Always write down at least 4 figures of the calculator display, and THEN round your answer.

The (radius) of a circle is 4 cm.
Work out the (area) of the circle.
Give your answer correct to the nearest cm². *(1388 January 2005)*

The radius is given. The area is required. So use $A = \pi r^2$

$$A = \pi \times 4^2$$
$$= \pi \times 16$$
$$= 50.26...$$
$$\text{Area} = 50 \text{ cm}^2 \text{ (to the nearest cm}^2\text{)}$$

WARNING ⚠️

4^2 means 4×4 **not** 4×2

Substitute $r = 4$ into $A = \pi r^2$

To find the value of $\pi \times 4^2$ on your calculator, you can just key in
$\pi \times 4 \; x^2 \; =$

EXAM TIP

Remember to write the units if the question asks you to.

Now try these

If your calculator does not have a π button, take the value of π to be 3.142
Give answers correct to 1 decimal place.

HINT

The **diameter** is given.

1 The diameter of a circle is 5.8 cm. Work out its circumference.

2 The radius of a circle is 13 mm. Work out its circumference.

HINT

The **radius** is given.

3 The radius of a circle is 4.7 cm. Work out its area.

4 The diameter of a circle is 9.72 m. Work out the area.

HINT

The diameter is given, but you need the radius to find the area.

5 The circumference of a circle is 19.7 cm. Work out its diameter.

6 The top of a table is a circle.
   The radius of the top of the table is 50 cm.
   a Work out the area of the top of the table.
   The base of the table is a circle.
   The diameter of the base of the table is 40 cm.
   b Work out the circumference of the base of the table.

(1387 June 2006)

HINT

Substitute $C = 19.7$ into $C = \pi d$

7 The diameter of a motorcycle wheel is 65 cm.
   a Work out the distance, in metres, the motorcycle travels when the wheel makes 100 complete turns.
   b Work out the number of complete turns the wheel makes when the motorcycle travels 500 m.

HINT

In one complete turn of the wheel, the distance travelled by the motorcycle is equal to the circumference of the wheel.

8 The circumference of a circle is 23.8 cm. Work out its area.

In Questions **9** and **10**, work out the areas of the shaded regions.

9

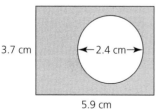

3.7 cm   ←2.4 cm→

5.9 cm

10

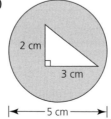

2 cm

3 cm

|← 5 cm →|

Diagrams **NOT** accurately drawn

HINT

First calculate the diameter or radius of the circle.

HINT

Use **subtraction** to find the areas of the shaded regions.

31 Semicircles and quarter circles

SKILL

Calculate the area and perimeter of a semicircle or a quarter circle

EXAM FACTS

Marks lost (%)	Marks available

KEY FACTS

circle   semicircle   quarter circle

- To find the area of a semicircle, divide the area of the full circle by 2
- To find the area of a quarter circle, divide the area of the full circle by 4
- To find the perimeter of a semicircle, divide the circumference of the full circle by 2 and add the diameter.
- To find the perimeter of a quarter circle, divide the circumference of the full circle by 4 and add 2 radii.

REFERENCE

For a reminder of how to find the circumference and area of a full circle, turn to pages 85 to 86

Getting it right

This working would score 1 mark.

Work out the perimeter of this quarter circle. Give your answer correct to 1 decimal place.

Diagram **NOT** accurately drawn

7.3 cm

Arc length $= \dfrac{2\pi r}{4}$

$= \dfrac{2 \times \pi \times 7.3}{4}$

$= 11.466...$

Perimeter $= 11.466... + (7.3 + 7.3)$

$= 11.466... + 14.6$

$= 26.066...$

$= 26.1 \text{ cm (to 1 d.p.)}$

WARNING ⚠️

This is only the arc length, not the final answer. The perimeter is the sum of the arc length **and** the two radii.

Adding on the 2 radii would also score 1 mark for method, even if you got the arc length wrong.

EXAM TIP

Write down at least 4 figures of the calculator display, before you round your answer.

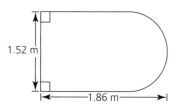

Diagram **NOT** accurately drawn

A mat is made in the shape of a rectangle with a semicircle added at one end.
The width of the mat is 1.52 metres.
The length of the mat is 1.86 metres.

Calculate the area of the mat.
Give your answer in square metres, correct to 2 decimal places.

(1385 November 1999)

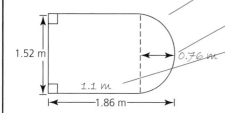

Split the shape into a rectangle and a semicircle. Just doing this scores 1 mark.

Radius of semicircle = 1.52 ÷ 2 = 0.76 m

Length of rectangle
= 1.86 – radius
= 1.86 – 0.76
= 1.1 m

To work out the area of the rectangle, multiply its length by its width.

Show how to work out the area of the semicircle. This will gain a mark.

Work out the area of the semicircle. Write down at least 4 figures.

Add the two areas.

Give the answer correct to 2 decimal places.

Area of rectangle = 1.1 × 1.52
$\qquad$ = 1.672 m²

Area of semicircle = $\frac{1}{2}$ × πr²

$\qquad$ = $\frac{\pi \times 0.76^2}{2}$

$\qquad$ = 0.9072... m²

Area of shape = 1.672... + 0.9072...

$\qquad$ = 2.579...

$\qquad$ = 2.58 m² (to 2 d.p.)

Now try these

If your calculator does not have a π button, take the value of π to be 3.142
Give answers correct to 1 decimal place.

In Questions **1–10**, for each shape, work out **a** the area **b** the perimeter.
In Questions **5–10** the rounded ends are semicircles or quarter circles.

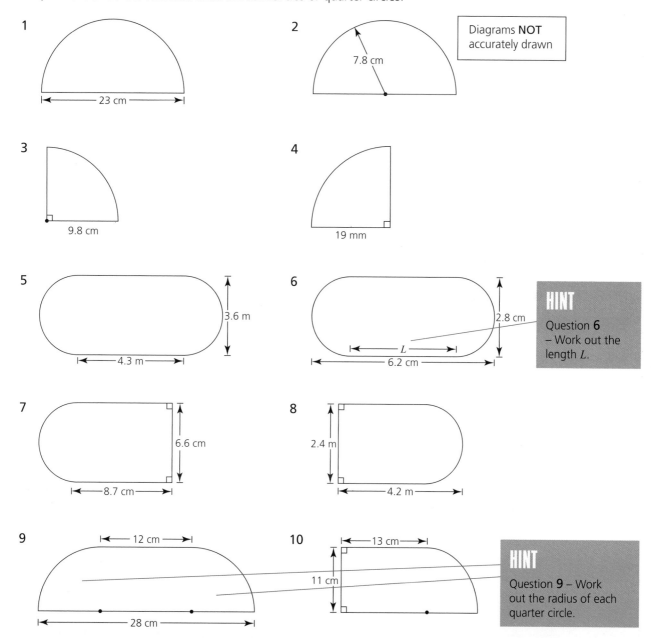

1

23 cm

2

7.8 cm

Diagrams **NOT** accurately drawn

3

9.8 cm

4

19 mm

5

3.6 m

4.3 m

6

2.8 cm

L

6.2 cm

HINT

Question **6** – Work out the length L.

7

6.6 cm

8.7 cm

8

2.4 m

4.2 m

9

12 cm

28 cm

10

13 cm

11 cm

HINT

Question **9** – Work out the radius of each quarter circle.

SKILL

Describe a transformation in words

EXAM FACTS

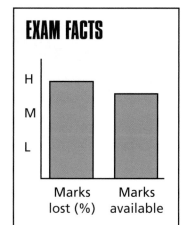

KEY FACTS

Transformation	Described by giving ...
• Translation	a vector such as $\begin{pmatrix} 5 \\ -4 \end{pmatrix}$ or how far right or left and how far up or down the shape moves.
• Rotation	the angle of turn, the direction of turn and the centre of rotation.
• Reflection	the mirror line.
• Enlargement	the scale factor and the centre of enlargement.

Getting it right

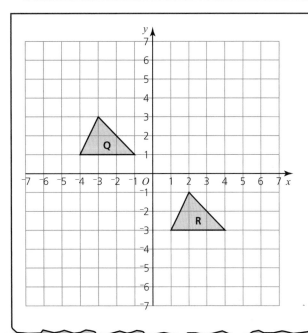

Describe fully the **single** transformation that takes triangle **Q** to triangle **R**.

(1388 November 2006)

When the question asks for a **single** transformation, no marks will be awarded for an answer which gives more than one transformation.

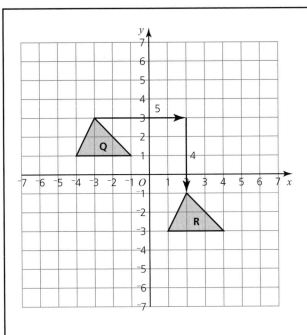

There are usually 2 marks for describing a translation, which suggests that 2 pieces of information must be given.

All points have moved 5 squares to the right and 4 squares down. So the transformation is a **translation**.

Translation with vector $\begin{pmatrix} 5 \\ -4 \end{pmatrix}$

or translation of 5 squares to the right and 4 squares down

State the type of transformation, the movement involved.

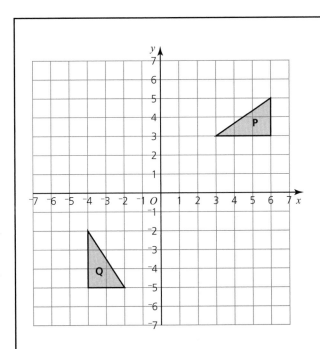

Describe fully the **single** transformation that takes triangle **P** to triangle **Q**.

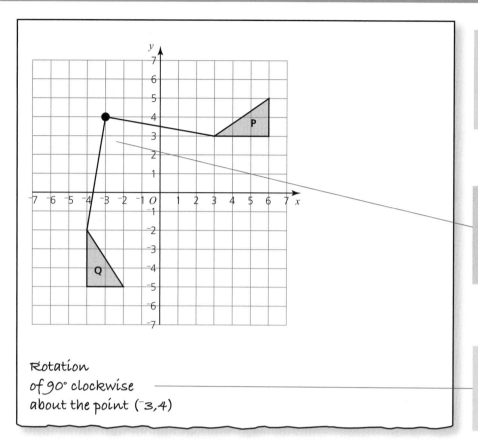

There are usually 3 marks for describing a rotation for which the centre of rotation is not the origin, which suggests that 3 pieces of information must be given.

WARNING

Errors are most likely with the centre of rotation.
Use tracing paper to find it.

Rotation
of 90° clockwise
about the point (⁻3,4)

State the type of transformation, the angle of turn and its direction, the centre of rotation.

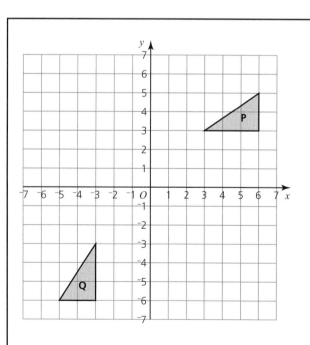

Describe fully the **single** transformation which maps triangle **P** onto triangle **Q**.

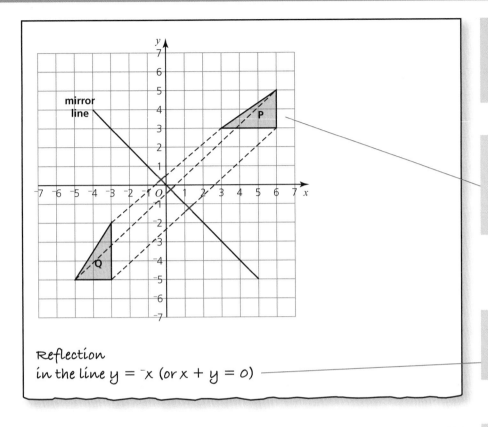

There are usually 2 marks for describing a reflection, which suggests that 2 pieces of information must be given.

Join each corner of triangle **P** to its image on triangle **Q**. The mirror line passes through the midpoints (marked with red crosses) of these lines.

State the type of transformation, the equation of the mirror line.

Reflection
in the line y = ⁻x (or x + y = 0)

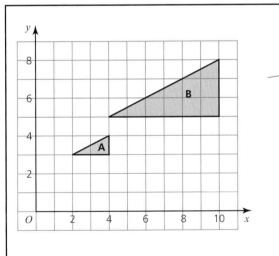

Triangle **B** is bigger than triangle **A** and so the transformation is an enlargement. Translation, rotation and reflection have no effect on the size of a shape.

Describe fully the **single** transformation which maps triangle **A** onto triangle **B**.

(4400 November 2006)

There are usually 3 marks for describing an enlargement for which the centre of enlargement is not the origin, which suggests that 3 pieces of information must be given.

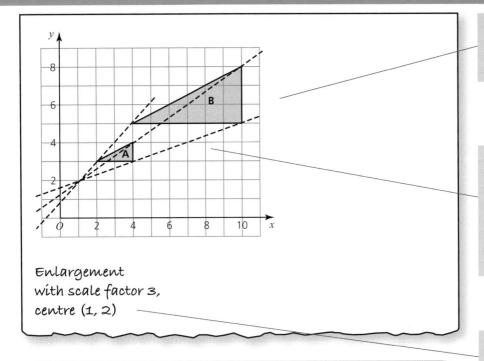

The length of each side of **B** is 3 times the length of the corresponding side of **A**. So the scale factor is 3

To find the centre of enlargement, join each corner of triangle **A** to its corresponding corner in triangle **B**. Extend each line until they meet at a point. This point is the centre of enlargement.

Enlargement
with scale factor 3,
centre (1, 2)

State
the type of transformation,
the scale factor,
the centre of enlargement.

Now try these

1   Describe fully the single transformation which takes triangle **P** to
   a triangle **Q**,
   b triangle **R**,
   c triangle **S**.

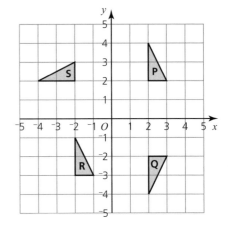

2   Describe fully the single transformation which maps triangle **P** onto
   a triangle **Q**,
   b triangle **R**,
   c triangle **S**,
   d triangle **T**.

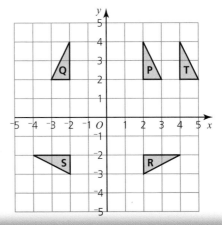

3   a   Describe fully the single transformation which maps
        triangle **T** onto
        triangle **U**.
    b   Describe fully the single transformation which maps
        triangle **U** onto triangle **T**.

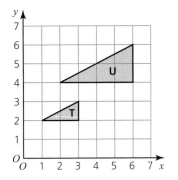

4   Describe fully the single transformation which takes
    triangle **P** to
    a   triangle **Q**,
    b   triangle **R**,
    c   triangle **S**,
    d   triangle **T**,
    e   triangle **U**,
    f   triangle **V**.

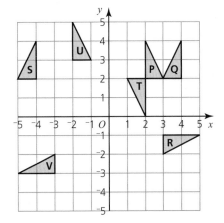

5   Describe fully the **single** transformation that maps
    triangle **A** onto triangle **B**.

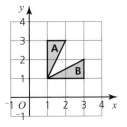

(1387 November 2006)

6   Describe fully the **single** transformation that takes
    triangle **P** to triangle **Q**.

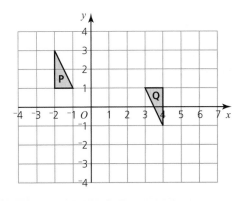

33 Pythagoras' theorem

SKILL

Use Pythagoras' theorem to find one side of a right-angled triangle, given the lengths of the other two sides

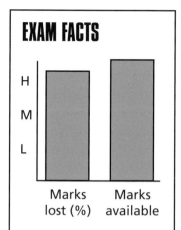
KEY FACTS

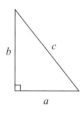

- In a right-angled triangle the side opposite the right angle (the longest side of the triangle) is called the **hypotenuse** of the triangle.

- In the diagram the length of the hypotenuse is c.

- Pythagoras' theorem states that the area of the square on the hypotenuse is equal to the sum of the areas of the squares on the other two sides.

- That is $c^2 = a^2 + b^2$

- In triangle DEF, Pythagoras' theorem gives $DE^2 = EF^2 + DF^2$

- DE^2 means that the length of the side DE is squared.

Getting it right

In triangle DEF
$FE = 8.7$ cm,
$DF = 6.4$ cm,
Angle $DFE = 90°$.
Calculate the length of DE.
Give your answer correct to 1 decimal place.

6.4 cm
8.7 cm

Diagram **NOT** accurately drawn

$DE^2 = EF^2 + DF^2$

$DE^2 = 8.7^2 + 6.4^2$

$DE^2 = 75.69 + 40.96$

EXAM TIP

"Diagram **NOT** accurately drawn" means that taking measurements from the diagram will not give the correct answer.

Identify the hypotenuse of the triangle (the side opposite the right-angle). Then write down Pythagoras' theorem for the triangle.

Substitute the given lengths. You would get 1 mark for this.

$DE^2 = 116.65$

$DE = \sqrt{116.65} = 10.80046\ldots$

$DE = 10.8$ cm

WARNING

A common error is to fail to find the square root and give the answer as 116.65

Remember to round your answer to 1 decimal place and write the units.

PQR is a right-angled triangle.
Angle $PQR = 90°$.
$QR = 15$ cm.
$PR = 19$ cm.
Work out the length of PQ.
Give your answer correct to 1 decimal place.

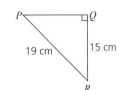

Diagram **NOT** accurately drawn

(1388 November 2005)

$$PR^2 = PQ^2 + QR^2$$
$$19^2 = PQ^2 + 15^2$$
$$361 = PQ^2 + 225$$
$$361 - 225 = PQ^2$$
$$PQ^2 = 136$$
$$PQ = \sqrt{136} = 11.6619\ldots$$
$$PQ = 11.7 \text{ cm}$$

WARNING

The side to be found is not opposite the right-angle so it is **NOT** the hypotenuse.
A common error is to write incorrectly $PQ^2 = 15^2 + 19^2$
This gives $PQ = 24.2$ which is not sensible, as PQ must be shorter than the hypotenuse, PR.

Now try these

In Questions **1–4**, work out the lengths of the sides marked with letters.
The diagrams are not accurately drawn.
Give each answer correct to 1 decimal place.

1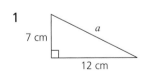

2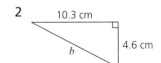

3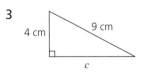

4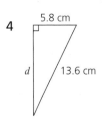

5 PQR is a right-angled triangle.
$PR = 6$ cm. $QR = 4$ cm
Work out the length of PQ.
Give your answer correct to 1 decimal place.

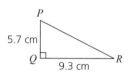

Diagram **NOT** accurately drawn

(1387 June 2006)

6 In triangle PQR
$QR = 9.3$ cm. $PQ = 5.7$ cm. Angle $PQR = 90°$.
Calculate the length of PR.
Give your answer correct to 1 decimal place.

Diagram **NOT** accurately drawn

(1388 November 2005)

7 Work out the value of x.

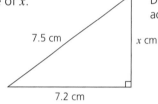

Diagram **NOT** accurately drawn

(4400 May 2006)

8 ABC is a triangle.
$AB = AC = 13$ cm.
$BC = 10$ cm.
M is the midpoint of BC.
Angle $AMC = 90°$.

Work out the length of AM.

Diagram **NOT** accurately drawn

(4400 November 2006)

9 The diagram shows three cities.
Norwich is 168 km due East of Leicester.
York is 157 km due North of Leicester.
Calculate the distance between Norwich and York.
Give your answer correct to the nearest kilometre.

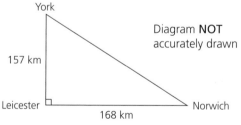

Diagram **NOT** accurately drawn

(1387 November 2006)

10 The diagram shows the positions of three
telephone masts A, B and C.

Mast C is 5 kilometres due East of Mast B.
Mast A is due North of Mast B and
8 kilometres from Mast C.

Calculate the distance of A from B.
Give your answer in kilometres, correct to
2 decimal places.

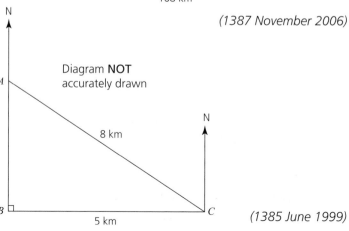

Diagram **NOT** accurately drawn

(1385 June 1999)

34 Averages and range

SKILLS

Find the mode
Find the median
Find the mean
Find the range

EXAM FACTS

H
M
L

Marks lost (%)   Marks available

KEY FACTS

- The mode is the number which occurs most times.
- The median is the middle number when the numbers are written in order.
- The mean = (sum of all the numbers) ÷ (how many numbers there are)
- The range is the difference between the highest number and the lowest number.

Getting it right

Here are the number of wheels on each of 12 vehicles.

2 4 6 4 3 2 8 4 5 8 5 6

a Find the mode.
b Find the median.
c Find the range.
d Find the mean.

> 4 occurs most times.

a 2 4 6 4 3 2 8 4 5 8 5 6

Mode = 4

> Write the numbers in order

b
1st	2nd	3rd	4th	5th	6th	7th	8th	9th	10th	11th	12th
2	2	3	4	4	4	5	5	6	6	8	8

↑
middle number

median = $\frac{4+5}{2}$ = 4.5

c Range = 8 − 2 = 6 — Range = highest number − lowest number

d 2+2+3+4+4+4+5+5+6+6+8+8 = 57

Mean = 57 ÷ 12 = 4.75

There are 12 numbers, so the median will be halfway between the 6th and 7th number. 4 and 5 are different so calculate their **mean**.

WARNING

A common error is to forget to order the list first and so write the answer as 5, the number halfway between 2 and 8 in the **unordered** list in the question.

WARNING

A common error for part **c** is an answer of "2 to 8".

> mean = (sum of all the numbers) ÷ (how many numbers there are)

WARNING

DO NOT round your answer 4.75 to 4.8

Rosie had 10 boxes of drawing pins.
She counted the number of drawing pins in each box.
The table gives information about her results.

Number of drawing pins	Frequency	
29	2	
30	5	
31	2	
32	1	

a Write down the modal number of drawing pins in a box.
b Work out the range of the number of drawing pins in a box.
c Work out the mean number of drawing pins in a box.

(1387 June 2003)

a Mode = 30
b Range = 32 − 29 = 3
c

Number of drawing pins	Frequency	Total
29	2	29 × 2 = 58
30	5	30 × 5 = 150
31	2	31 × 2 = 62
32	1	32 × 1 = 32

Total number of drawing pins = 58 + 150 + 62 + 32 = 302
Number of boxes = 10
Mean = 302 ÷ 10 = 30.2

5 of the 10 boxes have 30 drawing pins. 30 is the most common number of drawing pins in a box.

WARNING

A common error is to write 5 (the frequency) as the mode.

2 boxes each have 29 drawing pins
Total = 29 × 2 = 58
and so on.
This gets 1 mark.

"10 boxes" is given in the question so there is no need to add the frequencies.

WARNING

A common error is to find the mean of 29, 30, 31, 32

Now try these

1 Here are the test scores of seven students.

   1 3 3 7 8 9 11

a Write down the mode of the scores.
b Find the median score.
c Find the range of the scores.

(1388 November 2006)

2 Chloe made a list of her homework marks.

    4   5   5   5   4   3   2   1   4   5

a Write down the mode of her homework marks.
b Work out her mean homework mark.   *(1388 June 2005)*

3 Here are the heights of some fences in metres.

    1.2   2.3   1.7   1.9   2.6   2.3   1.2   1.4

a Find the median height.
b Find the range.
c Find the mean height.

4 Declan recorded the number of people in each of 20 cars.
His results are shown in the table.

Number of people	Frequency	
1	5	
2	4	
3	3	
4	7	
5	1	

a Find the range.
b Find the mean number of people in each car.

5 Habiba bought 25 packs of nails and counted the number of nails in each pack. The table shows her results.

a Write down modal number of nails in a pack.
b Find the mean number of nails in a pack.

Number of nails	Number of packs
16	10
17	6
18	0
19	7
20	2

6 The table shows the hourly pay, in pounds, of 120 workers.

Hourly pay (£)	Frequency	
8.50	29	
9.00	30	
9.50	29	
10.00	24	
10.50	8	

Find the mean hourly pay.

Stem and leaf diagrams

SKILLS

Draw a stem and leaf diagram

Use a stem and leaf diagram to find the median or the range of a set of data

KEY FACTS

- Stem and leaf diagrams are used to show data.
- From stem and leaf diagrams, the median and the range can be found easily.

EXAM FACTS

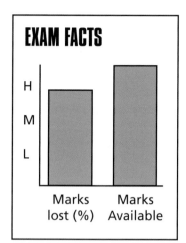

Getting it right

Here are the numbers of books borrowed from a school library on each of 12 days.

41 22 29 17 33 37 42 50 28 19 26 12

Draw an ordered stem and leaf diagram and key to show this information.

(1388 November 2006)

```
1 | 7 9 2
2 | 2 9 8 6
3 | 3 7
4 | 1 2
5 | 0
```

```
1 | 2 7 9
2 | 2 6 8 9
3 | 3 7
4 | 1 2
5 | 0
```

Key:
4 | 1 means 41

EXAM TIP

Even if the question does not mention it, the data must be ordered and a key must be included.
This question includes a reminder about this but many GCSE questions on this topic do not.

Record the numbers in the order in which they appear in the list, as this reduces the risk of missing any out. Count the number of entries in the table to make sure that all data has been included.

Order the units digits.

Any number in the list may be used as the key.

The key is important. You lose a mark if you omit it.

The stem and leaf diagram shows information about the pulse rate of each of 15 students.

```
5 | 6 8
6 | 0 2 3 8
7 | 1 4 6 6 8
8 | 7 8 9
9 | 7
```

Key:
5 | 6 means 56

a   Work out the range of the pulse rates.
b   Find the median pulse rate.

(1388 November 2005)

```
5 | (6) 8
6 | 0 2 3 8
7 | 1  4  6 6 8
8 | 7 8 9
9 | (7)
```

Key:
5 | 6 means 56

a   Lowest pulse rate = 56
    Highest pulse rate = 97
    Range = 97 − 56 = 41

> The range is the difference between the highest pulse rate and the lowest pulse rate, that is, the range is a single number.

b $\dfrac{15 + 1}{2} = \dfrac{16}{2} = 8$

> The middle of 15 numbers is the 8th number and so the median is the 8th pulse rate.

    Median = 74

Now try these

1   Jessica counted the number of words in each of the first 25 sentences of a book. Here are her results.

```
24   11   29   28   25   46   19   15   19   18   22   28   22
33   4    1    6    13   30   13   15   2    25   15   6
```

Draw an ordered stem and leaf diagram to show her results.
You should include a key.

(1387 November 2006)

HINT

Write 6 as 0 | 6

2 Daniel travels to school by bus. He recorded the time, in minutes, each day for fifteen days. His times are shown below.

21 18 24 31 21 30 19 22 24 32 33 28 22 29 18

Draw a stem and leaf diagram to show this information.

(1388 January 2003)

3 Fifteen players scored goals for Hawkshaw Rovers in the last three years. Here are the number of goals scored by each player.

10 8 13 24 25 33 10 1 21 16 3 24 16 31 20

Draw a stem and leaf diagram to show this information.

(1388 January 2005)

4 Here are the times, in minutes, taken to change some tyres.

5 10 15 12 8 7 20 35 24 15
20 33 15 25 10 8 10 20 16 10

Draw a stem and leaf diagram to show these times.

(1387 June 2003)

5 Mark recorded the number of e-mails he received each day for 21 days. The stem and leaf diagram shows this information.

Number of e-mails

0	4 5 5 6 7 7 8 9
1	0 1 2 3 3 4 6 7 8
2	0 1 3 6

Key:
2 | 6 means 26 e-mails

a Find the median number of e-mails that Mark received in the 21 days.
b Work out the range of the number of e-mails Mark received in the 21 days.

(1388 March 2005)

6 Shirin recorded the number of students late for school each day for 21 days. The stem and leaf diagram shows this information.

Number of students late

1	4 5 7 8 8 9
2	2 2 5 6 6 7 7 9 9
3	0 1 3 4 6

Key:
1 | 4 means 14 students late

a Find the median number of students late for school.
b Work out the range of the students late for school.

(1388 March 2003)

SKILLS

Find the modal class interval from a grouped frequency table

Find the class interval of grouped data which contains the median

KEY FACTS

- The modal class interval is the class interval with the highest frequency.

- The median is the middle value when all the data is placed in order. When the data is grouped the median cannot be found but the class interval that contains the median can be found.

- To find the position of the median,
  - work out $\dfrac{n+1}{2}$ where n is the total frequency
  - or find the class interval that contains the $\dfrac{n+1}{2}$ th piece of data.

WARNING

The most common error is to confuse mean, mode and median. Make sure that you learn the meaning of each term.

Getting it right

Fred did a survey of the time, in seconds, people spent in a queue at a supermarket. Information about the times is shown in the table.

Time (t seconds)	Frequency
$0 < t \leqslant 40$	8
$40 < t \leqslant 80$	12
$80 < t \leqslant 120$	14
$120 < t \leqslant 160$	16
$160 < t \leqslant 200$	10

Write down the modal class interval.

(1387 June 2006)

The highest frequency is 16
The modal class interval is $120 < t \leqslant 160$

Look in the frequency column to find the highest frequency.

WARNING

16 is the highest frequency, it is **not** the answer.

WARNING

A common error is to write down the mid-point value or the frequency as the answer. So, remember to write down the **class interval** as the final answer.

Bill recorded the times, in minutes, taken to complete his last 40 homeworks.

This table shows information about the times.

Time (t minutes)	Frequency	
$20 \leq t < 25$	8	
$25 \leq t < 30$	3	
$30 \leq t < 35$	7	
$35 \leq t < 40$	7	
$40 \leq t < 45$	15	

Find the class interval in which the median lies.   *(1387 June 2006)*

Time (t minutes)	Frequency	Cumulative Frequency
$20 \leq t < 25$	8	8
$25 \leq t < 30$	3	$8 + 3 = 11$
$30 \leq t < 35$	7	$11 + 7 = 18$
$35 \leq t < 40$	7	$18 + 7 = 25$
$40 \leq t < 45$	15	$25 + 15 = 40$

$$\frac{40 + 1}{2} = 20.5$$

The median is the 20.5^{th} time.
The median lies in the $35 \leq t < 40$ class interval.

WARNING

The most common error is to write down $30 < t \leq 35$ or 7. This is just the middle class interval, it does not necessarily contain the median.

Fill in the cumulative frequency. This is the same as a 'running total'.

1st to 8th times

9th to 11th times

12th to 18th times

19th to 25th times

Work out the position of the median, $\frac{n + 1}{2}$th position.

The median is in the middle of the 20th and 21st times. This will gain 1 mark.

WARNING

This gives the **position** of the median, it is **not** the answer.

The 20.5th time is in between the 20th and 21st times which are in the class interval that contains the 19th to 25th times.

WARNING

Remember to write down the **class interval** as your final answer.

Now try these

1   The grouped frequency table shows information about the number of hours worked by each of 200 headteachers in one week.

Number of hours worked (t)	Frequency
$0 \leq t < 30$	0
$30 \leq t < 40$	4
$40 \leq t < 50$	18
$50 \leq t < 60$	68
$60 \leq t < 70$	79
$70 \leq t < 80$	31

Write down the modal class interval.   *(1385 June 2001)*

2 Emma repairs bicycles. She keeps records of the cost of the repairs.
The table gives information about the costs of all repairs which she carried out in one week.

Find the class interval in which the median lies.

Cost ($£C$)	Frequency
$0 < C \le 10$	3
$10 < C \le 20$	7
$20 < C \le 30$	6
$30 < C \le 40$	8
$40 < C \le 50$	9

(1387 November 2005)

3 Jack grows apples.
He weighed 100 apples and recorded the weights to the nearest gram.
The table shows information about the weights (w grams) of the 100 apples.

a Write down the modal class.
b Find the class interval that contains the median.

Weight (w grams)	Frequency
$0 < w \le 60$	0
$60 < w \le 80$	16
$80 < w \le 120$	30
$120 < w \le 140$	25
$140 < w \le 160$	19
$160 < w \le 180$	10

4 70 office workers recorded the number of words per minute they could type.
The grouped frequency table gives information about the number of words per minute they could type.

a Write down the modal group.
b Find the class interval in which the median lies.

Number of words (w) per minute	Frequency
$0 < w \le 20$	10
$20 < w \le 40$	22
$40 < w \le 60$	21
$60 < w \le 80$	9
$80 < w \le 100$	5
$100 < w \le 120$	3

5 32 students took an English test.
There were 25 questions in the test.
The grouped frequency table gives information about the number of questions the students answered.

a Write down the modal class interval.
b Write down the interval which contains the median.

Number of test questions answered	Frequency
1 – 5	1
6 – 10	3
11 – 15	9
16 – 20	8
21 – 25	11

(1385 November 2002)

Estimating the mean of grouped data

SKILL

Use the midpoints in a grouped frequency table to find an estimate of the mean

EXAM FACTS

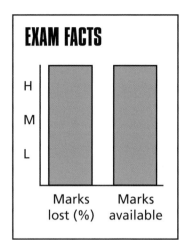

KEY FACTS

* To find an estimate for the mean from a grouped frequency table, use the middle value of each class interval.

Getting it right

The table shows some information about the number of minutes for which 120 students listened to music last Friday.

Number of minutes (t)	Frequency (f)		
$0 < t \leq 20$	16		
$20 < t \leq 40$	12		
$40 < t \leq 60$	20		
$60 < t \leq 80$	23		
$80 < t \leq 100$	30		
$100 < t \leq 120$	19		

Find an estimate for the mean number of minutes for which the students listened to music last Friday.

Number of minutes (t)	Frequency (f)	Middle value of the class interval (x)	$f \times x$
$0 < t \leq 20$	16	10	160
$20 < t \leq 40$	12	30	360
$40 < t \leq 60$	20	50	1000
$60 < t \leq 80$	23	70	1610
$80 < t \leq 100$	30	90	2700
$100 < t \leq 120$	19	110	2090

Estimate of total time = 7920

Total number of students = 120
Estimated mean = 7920 ÷ 120
= 66 minutes

WARNING

A common error is to divide the sum of the frequencies (120) by the number of class intervals (6). This gives an answer of 20 and gets no marks.

By using the middle values of the class intervals we are **estimating** the time spent by each student.

For example, to find the middle value of the class interval $40 < t \leq 60$, work out $\frac{1}{2}$ of $(40 + 60)$

Multiplying each middle value by the frequency gets 2 marks.

This is an **estimate** of the time spent listening to music by these 19 students.

Dividing the total of $f \times x$ by the total frequency, gets 1 mark.

Now try these

1 The table shows information about the weight of the luggage for each of 40 passengers on a plane.

Weight (w kg)	Frequency		
$0 < w \leq 4$	4		
$4 < w \leq 8$	10		
$8 < w \leq 12$	12		
$12 < w \leq 16$	9		
$16 < w \leq 20$	5		

Work out an estimate for the mean weight. *(1388 November 2006)*

HINT

Use the blank columns to work out x, middle value, and $f \times x$, frequency times middle value for each class.

2 75 boys took part in a darts competition.
Each boy threw darts until he hit the centre of the dartboard.
The numbers of darts thrown by the boys are grouped in this frequency table.

Number of darts thrown	Frequency		
1 to 5	10		
6 to 10	17		
11 to 15	12		
16 to 20	4		
21 to 25	12		
26 to 30	20		

Work out an estimate for the mean number of darts thrown by each boy.
(1385 November 2000)

3 The grouped frequency table shows information about the number of hours worked by each of 200 headteachers in one week.

Number of hours worked (t)	Frequency		
$0 < t \leq 30$	0		
$30 < t \leq 40$	4		
$40 < t \leq 50$	18		
$50 < t \leq 60$	68		
$60 < t \leq 70$	79		
$70 < t \leq 80$	31		

Work out an estimate of the mean number of hours worked by the headteachers that week.
(1385 June 2001)

4   35 students with Saturday jobs took part in a survey.
    They were asked the hourly rate of pay for their jobs.
    This information is shown in the grouped frequency table below.

Hourly rate of pay (£x)	Frequency		
$3.00 < x \le 3.50$	1		
$3.50 < x \le 4.00$	2		
$4.00 < x \le 4.50$	4		
$4.50 < x \le 5.00$	7		
$5.00 < x \le 5.50$	19		
$5.50 < x \le 6.00$	2		

Work out an estimate for the mean hourly rate of pay.
Give your answer to the nearest penny.

(1386 November 2002)

5   Charles found out the length of reign of each of 41 kings.
    He used the information to complete the grouped frequency table.

Length of reign (*L* years)	Number of kings		
$0 < L \le 10$	14		
$10 < L \le 20$	13		
$20 < L \le 30$	8		
$30 < L \le 40$	4		
$40 < L \le 50$	2		

Calculate an estimate for the mean length of reign.
Give your answer correct to 2 decimal places.

(1387 November 2003)

6   Fred did a survey on the areas of pictures in a newspaper.
    The table gives information about the areas.

Area (*A* cm^2)	Frequency		
$0 < A \le 10$	38		
$10 < A \le 25$	36		
$25 < A \le 40$	30		
$40 < A \le 60$	46		

Work out an estimate for the mean area of a picture.

(1387 November 2005)

38 Basic probability

SKILLS

Find probability using
successful outcomes ÷ total outcomes

Understand the meaning of 'mutually exclusive outcomes'

Find the probability that an outcome of an event will not happen

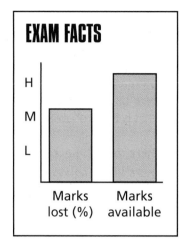
KEY FACTS

- Probabilities must be given as fractions, decimals or percentages.

- The probability of an outcome of an event is a measure of how likely it is that the outcome will happen.

- The probability of an outcome cannot be greater than 1

- When all possible outcomes are equally likely to happen,

$$\text{probability of an outcome} = \frac{\text{number of successful outcomes}}{\text{total number of possible outcomes}}$$

impossible $0 \leqslant$ probability $\leqslant 1$ certain

- Mutually exclusive outcomes are outcomes which cannot happen at the same time.

- The sum of the probabilities of all the possible mutually exclusive outcomes of an event is 1

- When the probability of an outcome happening is p, the probablity of the outcome **NOT** happening is $1 - p$.

Getting it right

Debbie is playing a game.
The probability that she will win the game is 0.59
Write down the probability that Debbie will **not** win the game.

(1388 January 2002)

P(win) = 0.59
P(not win) = 1 − 0.59
P(not win) = 0.41
The probability that Debbie will not win the game is 0.41

P(win) is a quick way of writing "the probability that Debbie wins".

"NOT A" means that A does not happen.
Use P(NOT A) = 1 − P(A)

Leave the answer as a decimal.

Helen has a bag of £1 coins.
5 of the coins are dated 2003
9 of the coins are dated 2004
6 of the coins are dated 2005
Helen takes at random one of the coins from the bag.
What is the probability that she will take a coin dated 2005?

There are 20 possible outcomes.
There are 6 coins dated 2005, that is 6 successful outcomes.

$$\text{probability of a 2005 coin} = \frac{\text{number of successful outcomes}}{\text{total number of possible outcomes}}$$

The probability that Helen will take a coin dated 2005 $= \frac{6}{20} = \frac{3}{10}$

Here is a 5-sided spinner.
Its sides are labelled 1, 2, 3, 4, 5
The spinner is biased.
The probability that the spinner will land on each of the numbers 1 to 4 is given in the table.

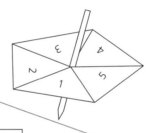

Number	1	2	3	4	5
Probability	0.36	0.1	0.25	0.15	

Alan spins the spinner once.
a Work out the probability that the spinner will land on 5
b Write down the probability that the spinner will land on 6

(1386 June 1999)

a The 5 possible outcomes, landing on 1, landing on 2, landing on 3, landing on 4, landing on 5, are mutually exclusive.

$P(5) = 1 - [P(1) + P(2) + P(3) + P(4)]$
$P(5) = 1 - [0.36 + 0.1 + 0.25 + 0.15]$
$P(5) = 1 - 0.86$
$P(5) = 0.14$

The probability that the spinner will land on 5 is 0.14

b The spinner does not have a 6 so it is impossible for the spinner to land on 6
The probability that the spinner will land on 6 is 0

1   Kevin buys a raffle ticket. A total of 350 raffle tickets are sold. One of these tickets will win the raffle. Each ticket has an equal chance of winning the raffle.
    Write down the probability that Kevin's ticket will win the raffle.
    (1388 March 2003)

2   Joshua rolls an ordinary dice once. It has faces marked 1, 2, 3, 4, 5 and 6
    Write down the probability that he gets
    a a 6,   **b** an odd number,   **c** a number less than 3,   **d** an 8
    (1387 June 2006)

3   The diagram shows a fair spinner in the shape of a regular octagon.
    The spinner can land on A or B or C.
    Marc spins the spinner.
    Write down the probability that the spinner will land on A.
    (1388 March 2005)

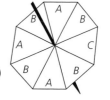

4   Asif has a box of 25 pens. 12 of the pens are blue. 8 of the pens are black. The rest of the pens are red. Asif chooses one pen at random from the box.
    What is the probability that Asif will choose
    a a blue pen,   **b** a red pen?   *(1386 June 2001)*

5   Bryan plays a game. The probability that he will win the game is 0.7
    Write down the probability that Bryan will **not** win the game.
    (1388 March 2005)

6   A bag contains coloured beads. A bead is selected at random.
    The probability of choosing a red bead is $\frac{5}{8}$. Write down the probability of choosing a bead that is **not** red from the bag.   *(1386 June 2000)*

7   The probability that it will snow in London on Christmas Day in any year is 0.08
    Work out the probability that it will **not** snow in London on Christmas Day.
    (1386 June 2002)

8   Four teams, City, Rovers, Town and United play a competition to win a cup. Only one team can win the cup. The table shows the probabilities of City or Rovers or Town winning the cup.
    Work out the value of x.

City	Rovers	Town	United
0.38	0.27	0.15	x

(1387 June 2003)

9   Imran plays a game of chess with a friend. A game of chess can be won or drawn or lost.
    The probability that Imran wins the game of chess is 0.3
    The probability that Imran draws the game of chess is 0.25
    Work out the probability that Imran loses the game of chess.
    (5540 June 2005)

10 Each day Uzma travels to work. She can be on time or early or late.
    The probability that she will be on time is 0.68
    The probability that she will be late is 0.07
    Work out the probability that Uzma will be early.   *(1388 November 2006)*

39 Expected frequency

EXAM FACTS

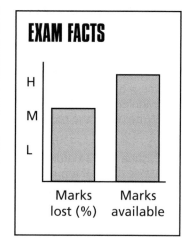

KEY FACTS

- When a statistical experiment is repeated many times, each time the experiment is carried out it is called a trial of the event.

- For each outcome of the event

$$\text{relative frequency} = \frac{\text{number of times the outcome happens}}{\text{total number of trials of the event}}$$

- It is best to give a relative frequency as a fraction. The value has to be between 0 and 1

- When the number of trials is large, the relative frequency of the outcome gives a good estimate for the probability of the outcome.

- When the exact probability of an outcome can be found, the value of the relative frequency gets closer to the exact probability, as the number of trials increases.

- When the probability of an outcome of the experiment is p and the experiment is carried out N times, then

an estimate for the number of times the outcome happens is $p \times N$

Getting it right

Tony throws a biased dice 100 times. The table shows his results.

Score	1	2	3	4	5	6
Frequency	12	13	17	10	18	30

He throws the dice once more.
a Find an estimate for the probability that he will get a 6.

Emma has a biased coin. The probability that the biased coin will land on a head is 0.7
Emma is going to throw the coin 250 times.
b Work out an estimate for the number of times the coin will land on a head.
(1387 November 2004)

> The instruction to "Find" or "Work out an estimate for…" does not mean that the answer is a guess. It is an estimate because the result of the experiment is uncertain.

a The dice was thrown 100 times and a 6 happened 30 times.

Estimate for the probability that he gets a 6 $= \dfrac{30}{100} = 0.3$

b Estimate for the number of times the coin will land on a head $= 0.7 \times 250$

$= 175$

Now try these

1 Mr Brown sows 200 flower seeds. For each flower seed the probability that it will produce a flower is 0.8
Work out an estimate for the number of these flower seeds that will produce a flower.

(1387 November 2006)

2 A dice has six faces numbered 1, 2, 3, 4, 5 and 6
The dice, which is biased, is thrown 200 times and the number on the upper face is recorded.
The frequencies of the numbers obtained are shown in the table.

Number shown on dice	1	2	3	4	5	6
Frequency	38	22	46	25	53	16

Work out an estimate for the probability that the next time the dice is thrown it will show the number 3

(1384 June 1997)

3 Saskia has a biased dice. The probability that the dice will land on a two is 0.6
Saskia is going to throw the dice 300 times. Work out an estimate for the number of times the dice will land on a two.

(1388 November 2006)

4 A solid cube has one red face and the other faces are coloured white.
The solid cube is biased. Sophie rolled the solid cube 200 times. The solid landed on the red face 46 times. The solid landed on a white face the other times. Sophie rolls the solid again.
Work out an estimate for the probability that the solid will land on a white face.

(1386 November 2001)

5 20 000 adults live in Mathstown. The probability that one of these adults, chosen at random, will vote in an election is 0.7
Work out an estimate for the number of these adults who will vote in an election.

(1388 March 2006)

6 The probability that a biased dice will land on a six is 0.4
Marie is going to throw the dice 400 times. Work out an estimate for the number of times the dice will land on a six.

(1388 January 2003)

7 Tony carries out a survey about the words in a book. He chooses a page at random. He then counts the number of letters in each of the first hundred words on the page.
The table shows Tony's results.

Number of letters in a word	1	2	3	4	5	6	7	8
Frequency	6	9	31	24	16	9	4	1

A word is chosen at random from the hundred words.
a What is the probability that the word will have 5 letters?
The book has 25 000 words.
b Estimate the number of 5 letter words in the book.

(1385 June 1999)

8 A computer game selects letters for players to use in the game. The table shows the probability of the letters A, E, I, O, U being chosen.

A	E	I	O	U
0.1	0.1	0.07	0.05	0.07

One letter is chosen at random.
a Find the probability of the letter U **not** being chosen.
300 letters are chosen for the game.
b Work out an estimate for the number of times the letter O will be chosen.

(1385 November 2002)

9 A bag contains counters which are white or green or red or yellow.
The probability of taking a counter of a particular colour at random is shown in the table.

Colour	White	Green	Red	Yellow
Probability	0.15	0.25		0.4

Laura is going to take a counter at random and then put it back in the bag.
a i Work out the probability that Laura will take a red counter.
ii Write down the probability that Laura will take a blue counter.
Laura is going to take a counter from the bag at random 100 times.
Each time she will put the counter back in the bag.
b Work out an estimate for the number of times that Laura will take a yellow counter.

(1385 November 2000)

10 Mark throws a fair coin. He gets a Head. Mark's sister then throws the same coin.
a What is the probability that she will get a Head?
Mark throws the coin 30 times.
b Explain why he may not get exactly 15 Heads and 15 Tails.

(1387 June 2004)

Key terms and exam vocabulary

Instruction	Meaning	Example	Notes
Write down	Little or no working is necessary.	Q The length of a line is 8 cm, correct to the nearest centimetre. Write down the **least** possible length of the line. A 7.5 cm	When "Write down" appears, the question is usually worth 1 mark.
Work out Calculate	These mean the same. Working or calculation is expected.	Q Work out an estimate for $\dfrac{496 \times 6.3}{0.48}$ A Estimate $= \dfrac{500 \times 6}{0.5} = \dfrac{3000}{0.5}$ $ = 3000 \times 2 = 6000$	There will usually be marks allocated to working.

Find	It could mean 'Write down' or it could mean 'Work out' or 'Calculate'.	Q Here is a list of numbers. 3 7 8 5 7
		  a Find the mode.
		  b Find the mean.
		
		A a Mode $= 7$
		  b Mean $= \dfrac{3 + 7 + 8 + 5 + 7}{5} = \dfrac{30}{5} = 6$
			Part **a** is "Write down" and part **b** is "Work out"

Simplify	These mean the same. "Fully" emphasises that the answer must be expressed as simply as possible.	Q Simplify $4x + 5y - x + 3y$ A $3x + 8y$	$3x + 5y + 3y$, for example, is also simpler but would not score full marks.
Simplify fully		Q Simplify fully $(5x^3y^2)^2$ A $25x^6y^4$	Used mainly in Higher tier algebra questions. Incomplete simplification such as $5^2x^6y^4$ would not score full marks.
Factorise	These mean the same. "Completely" warns the candidate against incomplete factorisation.	Q Factorise $x^2 - 8x$ A $x(x - 8)$	Only a single term can be taken outside the brackets.
Factorise completely		Q Factorise completely $6x^2 + 9x$ A $3x(2x + 3)$	Incomplete factorisation such as $3(2x^2 + 3x)$ would not score full marks.
Expand Multiply out	These mean the same.	Q Expand $5(x - 4)$ A $5x - 20$ Q Multiply out $5(x - 4)$	The answer cannot be simplified.
Expand and simplify	Like terms must be collected so that the answer is expressed as simply as possible.	Q Expand and simplify $(x + 4)(x - 1)$ A $x^2 - x + 4x - 4$ $ = x^2 + 3x - 4$	$x^2 - x + 4x - 4$ would not score full marks.

Solve	Find the number which the letter in an equation stands for.	**Q** Solve $3(x + 5) = 12$ **A** $3x + 15 = 12$ $\qquad 3x = 12 - 15$ $\qquad 3x = -3$ $\qquad x = -1$	$x = -1$ is called the **solution** of the equation. For full marks, $x = -1$ must be stated.
Construct	Use a ruler and compasses only to draw a shape. The ruler may be used to measure lengths, apart from the following four 'straight edge and compasses' constructions • an equilateral triangle with a given base • a hexagon inside a circle • perpendicular bisector of a given line • bisector of a given angle	**Q** The lengths of the sides of a triangle are 4.3 cm, 3.8 cm and 2.9 cm. Use ruler and compasses to construct this triangle accurately. You must show all construction lines. **A** 3.8 cm 2.9 cm 4.3 cm **Q** A point moves so that it is always equidistant from these two fixed lines. Construct its locus. **A**	Construction lines and arcs must be clearly visible. Don't rub them out. When drawing lines, you are expected to be within 1 mm of the required length.

Make a sketch	In a sketch, accurate lengths and angles are not required but it must be clear and as realistic as possible.	**Q** *ABCD* is a parallelogram. *AB* = 4.2 cm, *BC* = 3.9 cm and angle *ABC* = 141°. **a** Make a sketch of the parallelogram *ABCD*. **b** Make an accurate drawing of the parallelogram *ABCD*. **A a**	Use a ruler to draw the straight lines in a sketch. Notice that angle *ABC* is drawn as an obtuse angle.
Make an accurate drawing	Use a ruler, compasses and a protractor to draw a shape accurately.	**b**	Construction lines should be visible. When drawing lines, you are expected to be within 1 mm of the required length. When drawing angles, you are expected to be within 2° of the required size.

Answers

1 Writing a number as a product of its prime factors

1 $2 \times 2 \times 2 \times 5$
2 $3 \times 3 \times 5$
3 $2 \times 2 \times 13$
4 $2 \times 3 \times 3 \times 3$
5 $2 \times 2 \times 2 \times 3 \times 3$
6 $2 \times 2 \times 2 \times 2 \times 2 \times 3$
7 $2 \times 2 \times 5 \times 7$
8 $2 \times 2 \times 3 \times 3 \times 3$
9 $2^3 \times 5^2$
10 $2^3 \times 5^3$

2 Highest common factor

1 12
2 20
3 1, 2, 3, 4, 6, 12
4 1, 2, 3, 6, 9, 18
5 1, 2, 4, 5, 10, 20
6 1, 2, 3, 4, 6, 9, 12, 18, 36
7 1, 2, 3, 4, 6, 8, 12, 16, 24, 32, 48, 96
8 6
9 12
10 14
11 12
12 24
13 2
14 4
15 3
16 6
17 3
18 Jonathon is correct. Any even number must have 2 as a factor.
19 Becky is not correct. For example, the highest common factor of 8 and 12 is 4, not 2
20 a e.g. 15 and 45
    b No; even numbers which have 15 as a common factor are 30, 60, 90, 120, … The HCF of any pair of these is at least 30.

3 Lowest common multiple

1 3, 6, 9, 12, 15
2 4, 8, 12, 16, 20
3 7, 14, 21, 28, 35
4 11, 22, 33, 44, 55
5 13, 26, 39, 52, 65
6 8
7 36
8 30
9 24
10 48
11 60
12 96
13 30
14 12
15 24
16 30
17 36
18 Ben is wrong. 24 is a common multiple, but 12 is the lowest common multiple.

4 Long multiplication and multiplication of decimals

1 1302
2 4704
3 2976
4 12864
5 £1485
6 109.2
7 20.16
8 855.4 kg
9 £458.40
10 a 22.5   b 0.09
    c 25.92   d 2.03
    e 0.42   f 2.52

5 Long division and division of decimals

1 a 41   b 46
   c 35   d 26
2 a £51   b £13.60
   c £31.25   d £31.50
3 £0.18
4 £14.50
5 480
6 42
7 30.5 or $30\frac{1}{2}$ or $30\frac{13}{26}$ or 30 remainder 13
8 a 40   b 6
   c 0.4   d 42.1
   e 16   f 90
   g 3.6   h 24.4
9 a 310   b 31.2
   c 250   d 265
   e 188   f 70
   g 0.8   h 5.4
10 a 54   b 23
   c 4.3   d 370

6 Fractions of quantities

1. a 4 b 10
   c 7 d 26
   e 12 f 36
   g 25 h 28
2. 24
3. 8
4. 111
5. 36
6. 45 rooms
7. 72
8. £8.39
9. £105
10. £45
11. £75
12. 87 miles
13. 16
14. 75

7 Percentages of quantities

1. a 200 b 6
   c 2 d 6
   e 30 f 450
2. a £80 b 15 g
   c 2p d 12 kg
   e £30 f 270 cm
3. 520
4. 3180 kg
5. a £19.20 b £7.20
   c 6.57 m d 3.04 kg
   e £27.45 f 2.1 cm
6. £14.72
7. £33.84
8. 104
9. 720 000
10. 26
11. 540
12. £205

8 Percentage change and VAT

1. £1610
2. £144
3. £564
4. 76 kg
5. £58.75
6. £68
7. £32.80
8. £36 040
9. £493.50
10. £75.68
11. £2770.56
12. £6370
13. £0.42 or 42p
14. £75.20
15. £11 844
16. £0.18 or 18p

9 Using a calculator

1. 8.77
2. 5.062884553
3. 0.519434629
4. 1.57
5. 9.168
6. 4.232
7. 3.5
8. 0.119199815
9. 17.9867
10. 0.532329983

10 Sharing a quantity in a given ratio

1. a £18, £27 b £16, £20
   c £20, £120 d £20, £30, £50
   e £50, £100, £200
2. £21
3. £10.50, £14
4. £714
5. £78.75
6. 20 teachers
7. £180, £120, £420
8. £306
9. 8 sweets, 12 sweets, 16 sweets
10. 45 litres

11 Currency exchange rates

1. 2276 Rand
2. 560 euros
3. £81.08
4. £74.07
5. $336.60
6. a 395 euros b £52.63
7. a $1236 b £77.67
8. a 486 Swiss francs b £30
9. America by £15 or $27
10. £13.75
11. £20.37
12. £25

12 Estimates

1. 600
2. 5
3. 200
4. 10
5. 1.5
6. 60
7. 160
8. 3600
9. 2000 miles
10. £64

13 Basic algebra

1 a $6m$ b $2n^2$
   c $3pq$
2 a abc b $2de$
   c $18xy$
3 a $9xy$ b $-mn$
4 a $3e + 2f$ b $4g + 7h$
5 $7p - 8q$
6 $2a + 7b + 8$
7 a $x = 4$ b $y = \frac{16}{3} = 5\frac{1}{3}$
8 $x = 11$
9 $x = -0.5$
10 $y = -1.4$

14 Deriving expressions and formulae

1 £$(8S + 5b)$
2 $60 - m - p$
3 a pn b $\frac{25}{pn}$
4 $A = Pt + 23$
5 $x = 179 - 3y$
6 $T = 5p + 3q - r$
7 $P = 5x + 2y$
8 $x = \frac{bc}{p}$
9 $H = 50 - \left(\frac{60r + 47c}{100}\right)$ or $H = 50 - 0.60r - 0.47c$

15 Multiplying out brackets

1 a $3x + 3$ b $10 - 5y$
2 a $ab + 4a$ b $cd - ce$
3 a $20p - 30$ b $49 - 28q$
4 a $10xy - 2x$ b $3a + 9ab$
5 a $x^2 - x$ b $3y - 2y^2$
6 $6a + 7$
7 $3 - 10b$
8 $9a + 7b$
9 $8c - d$
10 $11m - 16$
11 $-2x - 14y$
12 $12z - 21$
13 $x^2 + 8x + 15$
14 $x^2 + x - 2$
15 $y^2 + 4y - 21$
16 $z^2 - 5z - 24$
17 $p^2 - 9p + 20$
18 $6 + 5x + x^2$
19 $y^2 + 2y + 1$
20 $x^2 - 8x + 16$

16 Negative numbers, expressions and formulae

1 7
2 -2
3 -6
4 a -6 b 0
   c 36
5 Uzma. $3n^2 = 3 \times 25 = 75$
6 a 9 b 100
7 7
8 19
9 a $-10°C$ b $-50°C$
10 -240

17 Tables of values

1

x	-3	-2	-1	0	1	2	3
y	5	6	7	8	9	10	11

2

x	-2	-1	0	1	2	3	4
y	19	14	9	4	-1	-6	-11

3

x	-3	-2	-1	0	1	2	3
y	15	10	7	6	7	10	15

4

x	-4	-3	-2	-1	0	1	2
y	64	43	26	13	4	-1	-2

5

x	-2	-1	0	1	2	3	4
y	-17	-1	7	7	-1	-17	-41

18 Graphs

1 a

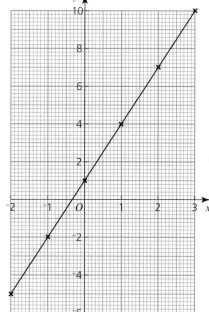

   b i $y = -1.4$
      ii $x = 2.4$

2 a

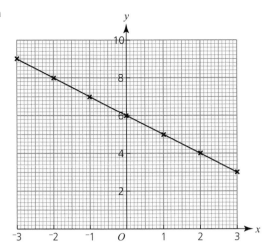

b $x = -1.4$

3 a

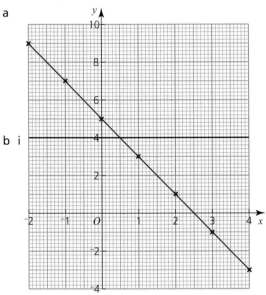

b i

ii (0.5, 4)

4 a

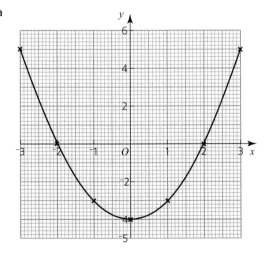

b minimum value of $y = -4$

5 a

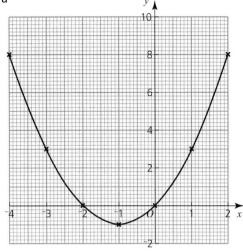

b $x = -3.24$ (accept any value from −3.3 to −3.2)

$x = 1.24$ (accept any value from 1.2 to 1.3)

6 a

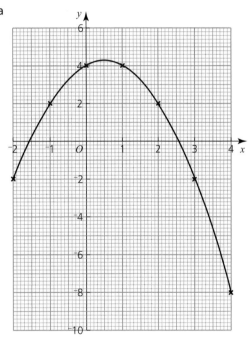

b 4.25 (accept 4.1 to 4.4)

c $x = -1.7$ (accept any value from −1.8 to −1.6)

$x = 2.7$ (accept any value from 2.6 to 2.8)

19 Solving equations that have brackets

1 $x = 4$

2 $y = 0.2$

3 $x = -21$

4 $p = 0.75$

5 $x = 18$

6 $x = -12.5$

7 $x = -\frac{7}{3} = -2\frac{1}{3}$

8 $x = 4$

9 $x = 0.6$
10 $x = 1$
11 $x = 0.8$
12 $y = -0.25$
13 $y = -8$
14 $a = \frac{6}{7}$
15 $p = \frac{5}{3} = 1\frac{2}{3}$

20 Setting up equations

1 a $6x + 6 = 24$ b 8 cm
2 a $5x - 20 = 180$ b $100°$
3 a $4x + 10$ b $x = 6$
4 a $4m - 6 = 2m + 5$ b $m = 5.5$
    c 16 kg
5 a $9r + 2$ b $r = 7$
6 a 18   b $35°$

21 Changing the subject of a formula

1 $c = \frac{f + 4}{3}$

2 $p = \frac{c + 2}{5}$

3 $r = \frac{3 - p}{4}$

4 $b = \frac{P - 2a}{2}$

5 $b = \frac{d - ac}{a}$ or $b = \frac{d}{a} = c$

6 $y = \frac{6x - 4z}{x}$ or $y = 6 - \frac{4z}{x}$

7 $m = \frac{y + 4}{5}$ or $B = \frac{CD}{A} = -2$

8 $r = 5P - 3$

9 $B = \frac{CD - 2A}{A}$ or $B = \frac{CD}{A} = -2$

10 a $x = \frac{2}{y}$ b $y = \frac{2}{x}$

22 nth term of an arithmetic sequence

1 $3n + 5$
2 22; $4n + 2$
3 a 31   b $4n - 1$
4 a 47   b $5n - 3$
    c 2, 4, 6
5 $22 - 5n$
6 a $7n - 21$ or $7(n - 3)$
    b Yes since 756 is a multiple of 7

23 Trial and improvement

2 5.3
3 3.6
4 b 3.6
5 4.4
6 3.8
7 2.7

24 Area of a triangle

1 9 cm²
2 20 cm²
3 60 m²
4 42 cm²
5 27.405 cm²
6 52.275 cm²
7 714 mm²
8 27.435 cm²
9 a 756 mm²   b 7.56 cm²
10 a 30 cm²   b 4.62 cm

25 Area and perimeter of compound shapes

1 63 m²
2 12 m²
3 66 m²
4 50 m
5 56 cm²
6 a trapezium   b 8.745 m²
    c 21.6 m²
7 45 cm²
8 54 cm
9 area of car park = 7080 m² so offer must be
   at least £198 240
   Mrs Roberts does NOT accept Mr Patel's offer.

26 Angle sums of triangles and quadrilaterals

1 $61°$
2 $128°$
3 $38°$
4 $66°$
5 $e = 32°$ $f = 74°$
6 $77°$
7 $113°$
8 $116°$
9 $31°$
10 a $x = 30°$ b $y = 48°$
11 a $k = 118$ b $l = 48$

27 Regular polygons – interior and exterior angles

1 $40°$
2 $72°$
3 $12°$
4 $36°$
5 a $60°$ b $120°$
6 $150°$
7 $170°$
8 a 24   b 40
9 15
10 18
11 $3240°$

28 Bearings

(Allow a tolerance of 2° for all bearings.)

1 a 016°   b 190°
  c 175°   d 277°
2 a Stoke-on-Trent   b Preston
  c Blackpool   d Chester
3 142°
4 290°
5

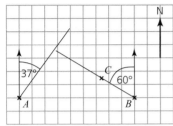

6

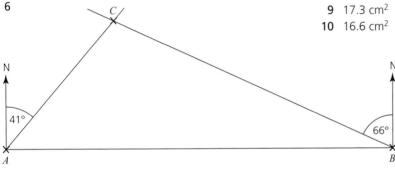

7 a 075°
  b

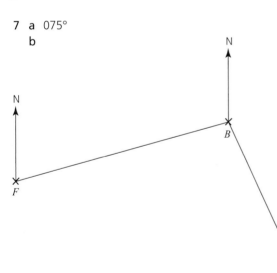

29 Speed

1 300 km/h
2 100 km
3 98 km/h

4 5.5 hours
5 26 km/h
6 a 126 km   b 2 hours 30 minutes
7 48 mph = 48 miles/hour
8 11 20
9 45 km/h
10 48.9 km/h

30 Circumference and area of a circle

1 18.2 cm
2 81.7 mm
3 69.4 cm²
4 74.2 m²
5 6.3 cm
6 a 7854.0 cm²   b 125.7 cm
7 a 204.2 m   b 244
8 45.1 cm²
9 17.3 cm²
10 16.6 cm²

31 Semicircles and quarter circles

1 a 207.7 cm²   b 59.1 cm
2 a 95.6 cm²   b 40.1 cm
3 a 75.4 cm²   b 35.0 cm
4 a 283.5 mm²   b 67.8 mm
5 a 25.7 m²   b 19.9 m
6 a 15.7 cm²   b 15.6 cm
7 a 74.5 cm²   b 34.4 cm
8 a 9.5 m²   b 12.2 m
9 a 96.5 cm²   b 65.1 cm
10 a 238.0 cm²   b 65.3 cm

32 Transformations

(0, 0) may be given instead of O and an angle may be given as a fraction of a turn instead of in degrees.

1 a reflection in x-axis
  b translation with vector $\begin{pmatrix} -4 \\ -5 \end{pmatrix}$
  c rotation of 90° anticlockwise about O
2 a reflection in y-axis
  b rotation of 90° clockwise about O
  c reflection in the line $y = -x$
  d translation with vector $\begin{pmatrix} 2 \\ 0 \end{pmatrix}$
3 a enlargement with scale factor 2, centre O
  b enlargement with scale factor $\frac{1}{2}$, centre O

4 **a** reflection in the line $x = 3$
 b rotation of 90° clockwise about (1, 0)
 c reflection in the line $x = -1$
 d rotation of 180° about (2, 2)
 e translation with vector $\binom{-4}{1}$
 f rotation of 90° anticlockwise about (2, −3)
5 reflection in the line $y = x$
6 rotation of 180° about (1, 1)

33 Pythagoras' theorem

1	13.9 cm	**2**	11.3 cm
3	8.1 cm	**4**	12.3 cm
5	7.2 cm	**6**	10.9 cm
7	2.1	**8**	12 cm
9	230 km	**10**	6.24 km

34 Averages and range

1 **a** 3   **b** 7   **c** 10
2 **a** 5   **b** 3.8
3 **a** 1.8 m   **b** 1.4 m   **c** 1.825 m
4 **a** 4   **b** 2.75
5 **a** 16   **b** 17.4
6 £9.30

35 Stem and leaf diagrams

1 Number of words

0	1	2	4	6	6			
1	1	3	3	5	5	5	8	9	9
2	2	2	4	5	5	8	8	9
3	0	3						
4	6							

Key: 2 | 4 means 24 words

2 Time (minutes)

1	8	8	9					
2	1	1	2	2	4	4	8	9
3	0	1	2	3				

Key: 2 | 1 means 21 minutes

3 Number of goals

0	1	3	8		
1	0	0	3	6	6
2	0	1	4	4	5
3	1	3			

Key: 1 | 0 means 10 goals

4 Time (minutes)

0	5	7	8	8			
1	0	0	0	0	2	5	5	5	6
2	0	0	0	4	5		
3	3	5					

Key: 0 | 5 means 5 minutes

5 **a** 12   **b** 22
6 **a** 26   **b** 22

36 Using grouped frequency tables

1 $60 \leqslant t < 70$
2 $30 < C \leqslant 40$
3 **a** $80 < w \leqslant 120$ **b** $120 < w \leqslant 140$
4 **a** $20 < w \leqslant 40$ **b** $40 < w \leqslant 60$
5 **a** 21 – 25   **b** 16 – 20

37 Estimating the mean of grouped data

1 10.1 kg
2 16.4
3 60.75 hours
4 £4.92
5 16.95 years
6 27.3 cm²

38 Basic probability

1 $\frac{1}{350}$

2 **a** $\frac{1}{6}$ **b** $\frac{3}{6} = \frac{1}{2}$
  c $\frac{2}{6} = \frac{1}{3}$ **d** 0

3 $\frac{3}{8}$

4 **a** $\frac{12}{25}$ **b** $\frac{5}{25} = \frac{1}{5}$

5 0.3

6 $\frac{3}{8}$

7 0.92

8 0.2

9 0.45

10 0.25

39 Expected frequency

1 160

2 $\frac{46}{200} = \frac{23}{100} = 0.23$

3 180

4 $\frac{154}{200} = \frac{77}{100} = 0.77$

5 14 000

6 160

7 **a** $\frac{16}{100} = \frac{4}{25} = 0.16$ **b** 4000

8 **a** 0.93   **b** 15

9 **a** **i** 0.2   **ii** 0
  b 40

10 **a** $\frac{1}{2} = 0.5$
  b Any explanation that indicates that any numbers of
   Heads and Tails from 0 to 30 are possible.

Index

Published by: Edexcel Limited, One90 High Holborn, London WC1V 7BH

Distributed by: Pearson Education Limited, Edinburgh Gate, Harlow, Essex CM20 2JE, England
www.longman.co.uk

© Edexcel Limited 2007

The rights of Trevor Johnson, Tony Clough, Julie Bolter, Michael Flowers, Rob Summerson and Kevin Tanner as the authors of this Work have been asserted by Trevor Johnson, Tony Clough, Julie Bolter, Michael Flowers, Rob Summerson and Kevin Tanner in accordance with the Copyright, Designs and Patents Act, 1988.

First published 2007
ISBN-13: 978-1-84690-187-4

Typeset by Pantek Arts Ltd
Printed in Great Britain by Scotprint Limited, Haddington

The publisher's policy is to use paper manufactured from sustainable forests.

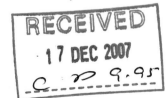